STRANGE THINGS

Clarendon Lectures in English Literature 1991

STRANGE THINGS

The Malevolent North in Canadian Literature

Margaret Atwood

CLARENDON PRESS . OXFORD
1995

Oxford University Press, Walton Street, Oxford OX2 6DP
Oxford New York
Athens Auckland Bangkok Bombay
Calcutta Cape Town Dar es Salaam Delhi
Florence Hong Kong Istanbul Karachi
Kuala Lumpur Madras Madrid Melbourne
Mexico City Nairobi Paris Singapore
Taipei Tokyo Toronto
and associated companies in
Berlin Ibadan

Oxford is a trade mark of Oxford University Press

Published in the United States
by Oxford University Press Inc., New York

British Library Cataloguing in Publication Data
Data available

Library of Congress Cataloging in Publication Data
Data available
ISBN 0–19–811976–3

1 3 5 7 9 10 8 6 4 2

Set by Hope Services (Abingdon) Ltd.
Printed in Great Britain
on acid-free paper by
Bookcraft Ltd,
Midsomer Norton, Bath

ACKNOWLEDGEMENTS

THE lectures are printed as they were originally given, with any supplementary material added in footnotes. I would like to thank Kim Scott Walwyn of Oxford University Press, who lured me to Oxford in the first place; Andrew Lockett, also of Oxford University Press, who saw the book through the publication process; Ruth Atwood and Sarah Cooper, who work with me and without whose aid I would never find my toothbrush, let alone put together anything as complicated as an index, and Ramsay Cook, for services rendered; Professors Douglas Gray and Christopher Butler, who took such good care of me while I was at Oxford, although I was in danger of dying of a surfeit of whipped cream; Dr Jeri Johnson, who took me to Kelmscott Manor, site of cannibalisms of a different order; Katherine Lacey of Lady Margaret Hall, who kindly consented to sing the ballad 'Lord Franklin'; Alan Patten, of the Oxford University Canadian Society, where I appeared one damp evening; the nice porters who came with the keys in the middle of the night and addressed me as Madam even though I had managed somehow to lock myself out of my own bedroom, and would, if not rescued, have had to face the next day's events wearing nothing but a bedspread; and the sound crew of Oxford, who laboured long and hard to make sure I was more or less audible.

Giving these lectures was, on the whole, a terrifying experience, although enjoyable in retrospect. Anyway I made it through without the loss of any appendages; which is what people often say when they come back from the Canadian North itself.

CONTENTS

Introduction

Strange Things consists of four lectures which I delivered in the spring of 1991, at Oxford University, as part of the Clarendon Lecture Series in English Literature. This series is designed as a kind of half-way house between the non-specialist public and the ivory tower—between those who gobble up the literary comestibles, in other words, and those who inform them about the structure and nutritional content.

Most of those who have given lectures in this series have been working academics, that is, people who actually teach at universities and write scholarly texts. I am not one of these; rather, I am no longer one. The best I can do is to paraphrase the Mock Turtle and say, 'Once I was a *real* academic.' The tears I shed in this respect are oddly crocodilian.

Having agreed, in a moment of curiosity and rashness, to deliver these lectures, I got cold feet. Here was I, a non-scholar—and a Canadian non-scholar at that—presuming to address an audience that might contain not only some real scholars, but some real scholars from England. What did I have to say to such people about English literature that they would not already know? Precious little, but I had an ace up my sleeve: these people might know everything about Beowulf and Virginia Woolf and even Thomas Wolf, but I could bet a few wooden nutmegs on the fact that they probably wouldn't know much about that other literary animal, Black Wolf, not to mention Grey Owl. If I were to talk about Canadian literature in English, instead of English literature proper, I could count on my material being almost completely *terra incognita*, which meant that I would be able to stick in lots of quotes to make the lectures longer.

Although a few individual and stalwart Canadian writers have managed to thrash their way eastwards across the Atlantic—one might mention, for instance, Michael Ondaatje, Robertson Davies, Mordecai Richler, and Alice Munro—Canadian literature as a whole tends to be, to the English literary mind, what Canadian geography itself used to be: an unexplored and uninteresting wasteland, punctuated by a few rocks, bogs, and stumps. Note that I do not speak of the Scots, Welsh, or Irish, nor of the ordinary reader; however, for a certain kind of literary Englishperson, Canada—lacking the exoticism of Africa, the strange fauna of Australia, or the romance of India—still tends to occupy the bottom rung on the status ladder of ex-British colonies.

Knowing this, what then could I say? What, that is, that would occupy the attention of the sceptical and the bored-in-advance? For one wild moment I thought of pulling a Grey Owl impersonation, turning up in buckskin and a feather head-dress and beginning, 'I come in peace, brother,' which was how Grey Owl himself—that devious shape-changing Englishman—once greeted the King of England. He got away with it, too. The English have always liked what they used to call Red Indians, and I felt I would probably be more acceptable in that disguise than in an ordinary, tedious old dress.

But a greater inspiration soon struck. The English, I knew, were very fond of cannibalism. If I could put some of that in,

I was off on the right foot. And so it turned out; at the sherry party after the first lecture, I was treated to the spectacle of a number of Oxford academics nibbling hors-d'œuvres and delicately discussing the question of who they would be prepared to eat.

'I wouldn't eat anyone I *know*.'

'I would eat someone if they mixed the bits up so I didn't know what I was getting. I mean, I wouldn't want to think I was eating a toe.'

'I wouldn't eat the liver. I *hate* liver.'

As they say in the how-to books on public speaking, you need to know your audience.

These lectures, then, are about a few of the more *outré* menu items to be found in Canadian literature. The entire series is called *Strange Things*, which is a quote from a Robert W. Service poem about the Canadian North. The first lecture, called 'Concerning Franklin and his Gallant Crew', is about the mysterious and disastrous Franklin expedition of the nineteenth century—the intent of which was to discover the North-west Passage, though the result was death for all—and its later echoes in Canadian literature. Why has this story had such resonance, and what sort of resonance has it had? Why is being lost in the frozen North—and going crazy there—still alive and kicking as a Canadian theme, even though most Canadians now live in cities? How did the North come to be thought of as a frigid but sparkling fin de siècle *femme fatale*, who entices and hypnotizes male protagonists and leads them to their doom? (Do not despair: there is some cannibalism in this lecture, since it is now a proven fact that some of the expedition members ended up inside some of the others.)

The second lecture is entitled 'The Grey Owl Syndrome'. Grey Owl was a famous Native Canadian who, in the 1930s, almost single-handedly saved the beaver from being trapped to extinction in central Canada. He was also, in reality, an

Englishman named Archie Belaney, who wore his assumed mask successfully until his death. This lecture is partly about how some writers have attempted to see Native people, and partly about how some have attempted to *be* Native people. What accounts for this urge to claim kinship, and to see wilderness as salvation—in direct opposition to the Franklin story? (The cannibalism in this lecture is metaphorical, since we learn how Ernest Thompson Seton's Woodcraft Indian Movement got eaten by Teddy Roosevelt and Lord Baden-Powell, and turned into the Boy Scouts.)

The third lecture is called 'Eyes of Blood, Heart of Ice: The Wendigo' and is about the dreaded and overtly and voraciously cannibalistic snow-monster of the eastern boreal forest, which has appeared in many stories and poems, with many variations. What is a Wendigo, and what does it eat when it isn't eating you? And, even more importantly, how can you avoid turning into one?

The fourth lecture is called 'Linoleum Caves', a title suggested by a sentence in Alice Munro's *The Lives of Girls and Women*: 'People's lives, in Jubilee as elsewhere, were dull, simple, amazing and unfathomable—deep caves paved with kitchen linoleum.' I was intrigued by the contrast between the domestic linoleum and the natural and potentially dangerous cave, and in women-in-the-North stories there is often such a contrast—sometimes with the linoleum being the more treacherous feature. In this lecture I attempt to look at what happens when women writers choose the wilderness as a locale. What becomes of the body of imagery built up by male writers, in which a female Nature opposes a male protagonist? If the North is a cold *femme fatale*, enticing you to destruction, is it similarly female and similarly fatal when a woman character encounters it? It is possible to make love with a bear? And, if there's any cannibalism to be done, what happens when it's women doing the eating? This lecture talks about how some women writers have adapted the imagery and mystique of the North, both positive and negative, to their own complex and devious purposes.

While I was still in the process of giving these lectures, I was interviewed by a young man from Canada who was studying at Oxford. He told me that he had a friend—also Canadian—who was concerned about the subject-matter I was discussing. This friend felt that I should not be talking about the North, or the wilderness, or snow, or bears, or cannibalism, or any of that. He felt that these were things of the past, and that I would give the English a wrong idea about how most Canadians were spending their time these days. What then—I asked—did this young man think I should be discussing? 'The literature of urban life,' was the reply. I said I thought that the English had quite a lot of urban life themselves, and that they didn't need to hear about it from me. I failed to say that the right idea could often be right from a sociological point of view, but was not necessarily right from a literary one. Given a choice between a morning spent in the doughnut shop and a little cannibalism, which would you take—to read about, that is? Alice Munro of course could handily work in both—but as a rule?

These lectures, then, leave out much. They leave out the literature of urban life, for one thing. They also leave out what I actually said at the beginning of each lecture. I felt, with Grey Owl, that a certain amount of ceremonial dress was required, and took care to wear earrings appropriate to each occasion. For the Franklin lecture I wore my Baffin Island Inuit female skinning-knife earrings; 'Inuit female skinning-knife' means knife used by Inuit females for skinning other things. They were made of polished bone. The Grey Owl lecture was a little easier: fringed leather jackets are readily available, and to go with mine I also wore some fringed leather earrings.

The Wendigo lecture posed a challenge: it would have been

very bad luck to have worn a pair of miniature Wendigoes in the ears for the second lecture, even if such existed, so instead I wore some earrings depicting what Wendigoes eat when there are no human beings around: frogs. For this lecture I also dressed in Wendigo colours: white for the heart of ice, red for the bloodshot eyes, and black for the decaying teeth. For the fourth lecture, which dealt with northern-oriented women's writing in Canada, I looked for some miniature women—in the best of all worlds these would have been miniature women wearing parkas, mittens, and snowshoes—but none were available. I had to settle for abstraction, and came up with some ovoid aluminum shapes with squiggly things at the bottom. Matching one's earrings to one's lecture topics would not of course have been done by a respectable person, but, as I have pointed out, I am not an academic.

1

Concerning Franklin and his Gallant Crew

This is the first of four lectures which I will be delivering over the course of the next two weeks, barring sudden illnesses or hurricanes. These lectures are roughly grouped around certain image-clusters that have appeared and reappeared in Canadian literature, and which are connected with the Canadian North.

A great deal has been made, from time to time, of the search for 'the Canadian identity'; sometimes we are told that this item is simply something we have mislaid, like the car keys, and might find down behind the sofa if we are only diligent enough, whereas at other times we have been told that the object in question doesn't really exist and we are pursuing a phantom. Sometimes we are told that although we don't have one of these 'identities', we ought to, because other countries do. Those doing the telling are usually academics or newspaper columnists, who seem to be under the illusion that everyone knows what 'the British identity' or 'the French identity' is, and that these things are concrete and indivisible nouns—something you could put on a tourist brochure, like the Eiffel

Tower or a beefeater, rather than aggregates of a great number of different items, cultural artefacts, places, and memories. If all we want is a costume for the Miss Universe contest, there's always the Mounties.

But surely the search for the fabled Canadian identity is like a dog chasing its own tail. Round about and round about it goes, with the tail whisking out of sight; whereupon it proclaims the tail elusive, fragile, threatened, or absent. And yet, as everyone can plainly see, there is the tail, as firmly attached to the dog as ever, continuing to wag or on the contrary to droop, according to the climate—climate is very important in Canada—or the climate of opinion.

What, then, does everyone else—that is, those not obsessed with the hopelessness of the quest—plainly see? What we plainly see, when we look for this 'identity' anywhere else, is, on first sight, a collection of clichéd images—that is, images that have been repeated so often and absorbed so fully that they are instantly recognized. These clichéd images—or cliché-images, to coin a phrase—are usually based on fact or historical reality of some sort, and they need to exist before art or literature can play with them, that is, make variations on them, explore them more deeply, utilize their imaginative power—for they do have imaginative power—or turn them inside out. What art can't do is ignore them altogether; unless, of course, the artist chooses to play with someone else's cliché-images, as some do.

What do we mean by 'the North'?

Until you get to the North Pole, 'North', being a direction, is relative. 'The North' is thought of as a place, but it's a place with shifting boundaries. It's also a state of mind. It can mean 'wilderness' or 'frontier'. But we know—or think we know—what sorts of things go on there. In the Canadian North of popular image, the Mounties with their barking dog teams relentlessly pursue madmen through the snow, prospectors stumble raving out of the bush clutching their little bags of gold-dust, jolly voyageurs rollick in their canoes, Indians rescue hapless whites who get endlessly lost in the woods, wolves devour lone hunters, or not, as the case may be; Eskimoes . . . well, you get the picture. The picture is that such pictures, and many more, exist and are fully recognizable to their society, and get used by politicians, by picture postcards, by cartoonists, by contestants in the Miss Universe beauty pageant, and also by literary writers. The ends pursued by literature are more obscure, its pathways more oblique, but the chief features of the terrain—the signposts—have remained strangely constant although the values ascribed to them have varied considerably.

I realize that I am addressing a British audience, not a Canadian one, and I will try to avoid using specialized vocabulary, such as 'toe rubbers', 'blackfly', and 'Chinook', without clarification. But I'm aware also that there may be some closet Canadians lurking incognito in the audience—they're everywhere, after all, and it's hard for the uninitiated to tell them apart from Americans or shrubs—and I apologize to such persons in advance for telling them things they may already know, such as the answer to the question 'When was the War of 1812?'

My first lecture will introduce you to some motifs of the North through the disastrous Franklin expedition of the mid-nineteenth century. The second one will explore the curious and anxiety-ridden phenomenon of whites reinventing themselves as Native people, and seeing the wilderness not as where you go to die but as where you go to renew life. The third will trace the fortunes of the legendary and cannibalistic Wendigo—in the land of the Wendigoes, it's not who you know but who you eat. And the fourth will explore a subject that at least half the audience will probably have been expecting—that is, how Canadian *women* writers have bent these wilderness literary traditions to their own sometimes devious purposes.

But before I start talking about Franklin, his dire fate, and his literary life after death, I think it would be helpful for me to say in advance what these lectures are not.

First, they are not an attempt to prove to you that such a thing as Canadian Literature actually exists. If I'd been delivering them twenty or thirty years ago, they might have been. Back then, it was a standard witticism in some quarters—even in Canada, especially in Canada—to say that the term 'Canadian Literature' was an oxymoron. Poets wrote satirical poems about its shoddy and derivative state, including recipes for its concoction that included such ingredients as one beaver, two Mounties, a sprinkling of maple leaves, and so on. In 1972 I myself published a book called *Survival*, which was dedicated to the premiss that there really was a Canadian Literature, that it was not the same as either American or English literature, and that it reflected some disquieting as well as some enlightening things about the society that had produced it. This book was bought and read by a great many people, hundreds of whom wrote me letters saying they were glad to know there was a Canadian Literature, because their high-school teacher had told them there wasn't. I wouldn't have got the same kind of letters in 1920 or even 1930, when school readers were full of Canadian pieces which were later tossed out because they were not considered modern enough. And I doubt that I would get the same kinds of letters today, although you never know. High-school teachers still seem dedicated to the premiss that what the young mind really craves is *The Catcher in the Rye.*

Second, these lectures are not a survey of what is usually called 'the field'. No four lectures could be, unless they were to consist of nothing but titles. Nor do they attempt to be fair or all-inclusive in their coverage, even of their own topics. I am neither a scholar nor a specialist, and my choices are the result, not of extensive research, but of my own amateur enthusiasms. If you ask a writer to give a lecture, you'll get a writer's lecture; and as we all know, the insides of writers' heads resemble squirrels' nests more than they do neatly arranged filing-cabinets.

Third, I have not dealt with works written in French—not only because of the political ambiguities involved, but also for the simple reason that to do so would have kept you in your seats much longer than would be conscionable. The human rear can only endure so much reality. Fourth, these lectures are not about style, nor are they about individual authors and their *œuvres*; but neither do they take the point of view that individual authors can't be said to exist. (As an author, I've always had some difficulty with that.) They are not about rhetoric; they do not deconstruct anything, they do not problematize texts, and they will not leave you gazing vertiginously into the linguistic abyss or watching in alarm while the meaning of 'meaning' pops like bubble-gum and language vanishes up its own nether end.

These lectures depart from the position that, although in every culture many stories are told, only some are told and retold, and that these recurring stories bear examining. If such stories were parts of a symphony you'd call them leitmotifs, if they were personality traits you'd call them obsessions, and if it were your parents telling them at the dinner-table during your adolescence you'd call them boring. But, in literature, they hold a curious fascination both for those who tell them and for those who hear them; they are handed down and reworked, and story-tellers come back to them time and time again, approaching them from various angles and discovering new and different meanings each time the story, or a part of it, is given a fresh incarnation.

In Canadian literature, one such story is the Franklin expedition. For Americans, of course, the word *Franklin* means Benjamin, or else a stove. But for Canadians it means a disaster. Canadians are fond of a good disaster, especially if it has ice, water, or snow in it. You thought the national flag was about a leaf, didn't you? Look harder. It's where someone got axed in the snow.

The facts of the Franklin fiasco are fairly well known. In May of 1845 Sir John Franklin and 135 men, including the other Arctic veterans Captain Francis Crozier and Commander James Fitzjames, sailed from England on a voyage of discovery. Their two ships were named, with horrible prescience, the *Terror* and the *Erebus*. These ships had fortified hulls which were supposed to be able to withstand the tremendous pressure of pack-ice; they were provisioned for three years; and they contained many extras, such as hot water, steam heating, instruments for scientific research, and two libraries with a total of 2,900 volumes. For their time, they were the most technologically advanced and luxurious ships ever sent on such an expedition. They even had steam-driven screw propellers, a striking innovation for 1845.

Franklin's intent was to discover the North-west Passage, which Europeans had been trying to do without success for over 300 years. If charted, such a passage—in those days before the Panama Canal—would have made trade with China and India much faster, and therefore much more lucrative. The real goal of the expedition, then, was financial; but both the excited press of the time and later recountings have glorified it with other and loftier adjectives, of the *brave*, *heroic*, *gallant*, *daring*, and *selfless* variety. At the time, this venture was launched with unqualified optimism: the Franklin expedition, it was felt, would run no real risks and could not possibly fail, due to the wisdom of its leaders and, especially, to the up-to-date nature of its ships and supplies.

The expedition was last sighted in Baffin Bay, in July of 1845, by two whaling ships. After that the *Terror* and *Erebus* vanished, and none of their men were ever seen alive again. I say 'alive', because some of them *were* seen dead, under unusual circumstances.

When the expedition did not return, great efforts were made to find it. Several dozen other expeditions were sent out, and large rewards were offered—one by Lady Jane Franklin, Sir John's wife. In August of 1850 some graves dated the winter of 1846 were discovered on Beechey Island. They were the graves of three expedition members who must have died while still aboard ship: Braine, Hartnell, and Torrington. After that, a large cairn of discarded food tins was located—but no other clues, and no other corpses.

In 1854 the surveyor John Rae obtained second-hand news of the expedition's fate through some Inuit hunters, who reported that the ships had been crushed in the ice and the expedition's men had starved to death while trying to make their way south, indulging in a little cannibalism along the way. (Mention has been made, for instance, of a bootful of human flesh.[1]) When the subject was first brought up, the Inuit were indignantly denounced as barbaric liars, because a Briton could not possibly have behaved so hungrily; though later research has proved them true.[2] The Inuit had picked up various items from the expedition, including some monogrammed silver spoons, which Rae was able to purchase and display as proofs of his story. In its reporting, the Toronto *Globe* put a Tower of Babylon or Prometheus-the-Fire-Stealer spin on this story, turning it into a parable about overreaching yourself by foolishly trying to storm 'winter's citadel'.

In May 1859 a joint expedition headed by Captain Francis McClintock and William Hobson uncovered some harder evidence. Hobson discovered a message left in a stone cairn, which indicated that the ships had been deserted after having been 'beset'—that is, frozen in and unable to get out—for three years, and that the men had headed overland to try to

[1] From accounts by explorer Charles Francis Hall; see Owen Beattie and John Geiger, *Frozen in Time* (Saskatoon, 1989), 44 and 60.

[2] See e.g. the reports by physical anthropologist Anne Keenleyside in 'Bones of Contention' by Barry Ranford, *Equinox*, 74 (Spring 1994), 69.

locate the Back Fish River hundreds of miles to the south. By this time Franklin himself was already dead; the truth was—according to Native sources anyway—that the man was always somewhat of a fool, and had on previous occasions ignored local advice and gone places he'd been told not to. Possibly that was what had happened this time; coupled with a run of unusually cold weather, it spelled calamity. Meanwhile, McClintock had located some actual bodies, grouped in and around a ship's boat which these men had been trying to drag across land. The boat was filled with all kinds of useless junk—useless, that is, for basic survival: pocket combs, soap, slippers, toothbrushes, and a copy of *The Vicar of Wakefield.* The odd thing was the evidence of crazed behaviour: the boat was pointing the wrong way, and its provisioning was bizarre, to say the least. Whatever the men had been up to, it was obvious that they had not been thinking very clearly at the time.

In the early 1980s, a team of anthropologists, including University of Alberta forensic specialist Owen Beattie and his partner, Arctic archaeologist James Savelle, located the three sailors' graves and carefully exhumed Torrington, Braine, and Hartnell, who had been permafrosted ever since 1846. From samples of hair and fingernails they examined, the scientists concluded that these three men had been suffering from high levels of lead poisoning at the time of their deaths. In the 1840s the art of canning food had not yet been perfected. The seams of the cans were soldered with lead—the effects of which were not understood at the time—and the lead seeped into the food. The Franklin expedition, through its reliance on 'state-of-the-art' technology, had unwittingly poisoned itself. This would explain the disorientation of the men in their final days, and their apparent inability to devise and carry out the necessary survival strategies. If they'd been relying on more 'primitive' methods, such as hunting and fishing, they would have stood a much better chance. (Should you wish to know more about this remarkable piece of forensic deduction, read *Frozen in Time* by Owen Beattie and John Geiger, from which all of this information has been taken.)

These are the bare bones—as it were—of the Franklin story. But what has the Canadian literary imagination made of it?

Let's start with something that is not, in origin, Canadian at all, but English—the nineteenth-century ballad 'Lord Franklin' (though Franklin was not a lord, death seems to have promoted him)—or, in its many variations, 'Lady Franklin's Lament' or 'The Franklin Expedition'. This song is best known in England through the version sung by squeezebox-player A. L. Lloyd on the Ewan McColl record *Row, Bullies, Row.* There are several other versions still sung in Canada, with both different words and different music; the well-known Canadian song-writer Wade Hemsworth presents one of them, commenting, 'This song is meant to be sung solo in a country way. The balladeer stands up, supporting himself on the back of a chair usually, and roars it across the parlour in the old-fashioned way. . . . You find people down east who still do it—there are people in northern Ontario who will do it, and in Québec—but particularly down east. . . . They sing without accompaniment.'[3]

THE FRANKLIN EXPEDITION

The other night on the rolling deep
While in my hammock I lay asleep,
I dreamed a dream, and I thought it true,
Concerning Franklin and his gallant crew.

[3] Wade Hemsworth, *The Songs of Wade Hemsworth*, ed. Hugh Verrier (Waterloo, Ont., 1990), 132.

With a hundred seamen he sailed away,
To the frozen ocean in the month of May,
To seek that passage beyond the pole,
Where we poor seamen do sometimes go.

Through cruel hardships they vainly strove,
Their ships on mountains of ice were drove,
Where the Eskimo in his skin canoe
Is the only one who ever gets through.

In Baffin's Bay where the whale-fish blow,
The fate of Franklin no man may know;
The fate of Franklin no man can tell,
Lord Franklin along with his poor sailors do dwell.

And now my mem'ry it gives me pain,
For my long-lost Franklin I would cross the main;
Ten thousand pounds I would freely give
To know on earth that my Franklin do live.[4]

Beattie quotes the last two verses of this song in *Frozen in Time*, saying that the speaker is supposed to be Lady Franklin, who did indeed search long and hard for her vanished husband. Maybe the 'ten thousand pounds' point to her, but the narrator in the rest of the song is surely a man, a 'poor sailor' like the ones lost, and the identification is that of a man who can imagine himself in a similar plight—trapped in the frozen wastes, unable—unlike the local Eskimo inhabitants—to make his way through. The dream that is thought 'true' is the dream of finding what is lost, but by the end of the song that dream has proven false and the speaker is left with the ghostlike demi-existence, not only of Franklin, but of his entire expedition. As we know from other stories of mysterious vanishings at sea, those vanished have an odd quality of continued existence. Because Franklin was never really 'found', he continues to live on as a haunting presence; certainly in Canadian literature.

[4] Traditional.

However, the Franklin disaster did not take root in the Canadian imagination immediately; possibly because, at the time, the whole thing—enwrapped as it was in the Union Jack and in sentiments of the most grovelling patriotism—was thought of as too British. Although the rewards offered, and the findings of physical evidence, were given big space in the Toronto *Globe*, on at least one occasion the same paper complained that Tom Thumb the Midget's Toronto appearance vastly outdrew a public lecture on Franklin.[5]

But that was in the mid-nineteenth century. Later on, when Canada was beginning to identify its very own nightmares, the Franklin expedition was among them.

Before getting to literary treatments of the Franklin expedition proper, I'd like to mention the body of mystic-North imagery both built up and exploited by a turn-of-the-century poet whose work is usually ignored by serious critics—Robert W. Service. Service, of Sam McGee and Dan McGrew fame, wrote a great deal of Kiplingesque verse, much of which centred around his experiences in the Yukon during the Gold Rush. He was reflecting an already-existing body of lore and cliché—some of which was paralleled by actual travellers' tales—when he wrote poem after poem describing both the uncanny lure of the North and the awful things it could do to you, which included freezing you stiff and driving you crazy. Here are a couple of samples—the last verse of a poem called 'The Land God Forgot', for instance:

> O outcast land! O leper land!
>   Let the lone wolf-cry express
> The hate insensate of thy hand,
>   Thy heart's abysmal loneliness.[6]

Or, better, a snippet from 'Death in the Arctic', which is about a sole survivor wondering whether to shoot himself as he

[5] Owen and Geiger, *Frozen in Time*, 28.

[6] Robert W. Service, *The Complete Poems of Robert Service* (New York, 1945), p. xviii.

crouches in an igloo with the rest of his group, laid out in a gelid row in their sleeping-bags (why he didn't put them outside, like a sensible fellow, the poem does not say):

> Oh would you know how earth can be
> A hell—go north of Eighty-three![7]

Or how about 'The Law of the Yukon':

> Staggering blind through the storm-whirl, stumbling mad through the snow,
> Frozen stiff in the ice-pack, brittle and bent like a bow;
> Featureless, formless, forsaken, scented by wolves in their flight,
> Left for the wind to make music through ribs that are glittering white;
> Gnawing the black crust of failure, searching the pit of despair . . .[8]

. . . etc.

Service also has a lot of poems in which the North is endowed with beckoning voices; the title of his first collection is 'The Spell of the Yukon'—the Yukon is a 'she', and says things like,

> I am the land that listens, I am the land that broods;
> Steeped in eternal beauty, crystalline waters and woods.
> Long have I waited lonely, shunned as a thing accurst, . . .
> And I wait for the men who will win me . . .[9]

and so forth.

Service habitually personifies the North as a savage but fascinating female, and a talkative one at that. He's fond of titles such as 'The Call of the Wild' and 'The Lure of Little Voices'. These latter 'calling from the wilderness, the vast and God-like spaces, | The stark and sullen solitudes that sentinel the Pole', say to the narrator's wife—a rival female—'He was ours before you got him, and we want him once again.'[10]

A real-life footnote: in 1918 the painter Tom Thomson,

[7] Robert W. Service, *The Complete Poems of Robert Service* (New York, 1945), p.221.
[8] Ibid. 11. [9] Ibid. 12. [10] Ibid. 23.

known for his canvasses of the North, was found floating face-down beside his canoe. He was an experienced woodsman, and was in the habit of going off for long trips, by himself, to paint. There was no indication of how he had come to drown. But everyone knew, or thought they did: the Spirit of the North had claimed him as her own. The death of Tom Thomson was treated as somehow legendary, somehow exemplary. Now, a lot of people in Canada fall off their kitchen step-ladders and break their necks, and many others die in car crashes; and yet there is no mythology centring around these deaths, no Spirit of the Kitchen or Spirit of the Automobile that lures you on or claims you as its own. Significant deaths, like significant lives, are those we choose to find significant. Tom Thomson's death was found significant because it fitted in with preconceived notions of what a death in the North ought to be.

To sum up: popular lore, and popular literature, established early that the North was uncanny, awe-inspiring in an almost religious way, hostile to white men, but alluring; that it would lead you on and do you in; that it would drive you crazy, and, finally, would claim you for its own.

The first Canadian poet I know of who was fascinated by the Franklin story was E. J. Pratt. Pratt was a Newfoundlander who specialized in narrative poems dealing with epic ventures, in which—typically—groups of men struggle against odds imposed largely by an immense and difficult and usually very cold Nature. Writing between the 1920s and the 1950s, Pratt identified so many stories that were, and possibly still are, central to the Canadian literary imagination—including the building of the great railroads and the tale of the martyred French Jesuits of the seventeenth century—that you'd almost have to wonder what was wrong with the Franklin expedition if Pratt *hadn't* been interested in it.

But, as it turns out, he was. He made a couple of attempts to write about it and he had certainly read several accounts of it, such as Henty's *North Overland with Franklin* and Frank Shaw's *Famous Shipwrecks*, but he felt he couldn't undertake his planned Franklin epic without seeing the actual terrain for himself; and back in the 1930s it wasn't at all easy to get up to Baffin Island.[11]

In 1933 he was on the verge of starting his major Franklin poem; but, frustrated again in his attempts to head north, he tackled instead another saga of disaster—the sinking of the *Titanic.* It is the contention of this lecture that, in the hands of Pratt at least, the stories are very similar; that Pratt incorporated some of his thinking about the North and the Franklin catastrophe into his *Titanic* poem; and that, had he written the Franklin poem instead of the *Titanic* one, he would have given it much the same moral twist.

Pratt, like Shakespeare, gets the 'argument' out of the way at the beginning, in order to leave the decks clear for the action. (Or, rather, for the *in*action; as in the case of the Franklin expedition, there isn't a lot for the men to *do,* in the circumstances, except endure, and Pratt reserves his highest heroic points for those who hold back and let others scramble into the lifeboats. To paraphrase Milton, They also serve who only stand and sink.)

As in a history play, Pratt assumes that we already know the story: how the *Titanic* was thought to be unsinkable, how it was the most luxurious liner ever built, how, in its rash and heedless rush across the Atlantic, hoping to establish a record, it bumped into an iceberg, which opened it up like a can of sardines, and how a great many people subsequently drowned, some of them behaving well, others badly.

But what to make of this? Pratt had read Thomas Hardy's treatment of the *Titanic*—'The Convergence of the Twain'—with interest; but, in theory anyway, he repudiated Hardy's fatalistic vision. The *Titanic* disaster was not a case of the

[11] David G. Pitt, *E. J. Pratt: The Master Years, 1927–1964* (Toronto, 1987), 124–5.

Great Ironist arranging a predestined meeting between ship and iceberg, but of man's overweening arrogance. The second section of the poem, which deals with the *Titanic* as an engineering wonder, is a small hymn by man to himself, modelled on the press clippings of the time—and with Pratt's commentary on it added at the end.

> Completed! Waiting for her trial spin—
> Levers and telegraphs and valves within
> Her intercostal spaces ready to start
> The power pulsing through her lungs and heart.
>
> . . . .
>
> No storm could hurt that hull—the papers said so.
> The perfect ship at last—the first unsinkable,
> Proved in advance—had not the folders read so?
> Such was the steel strength of her double floors
> Along the whole length of the keel, and such
> The fine adjustment of the bulkhead doors
>
> . . . .
>
> That in collision with iceberg or rock
> Or passing ship she could survive the shock,
>
> . . . .
>
> And this belief had reached its climax when,
> Through wireless waves as yet unstaled by use,
> The wonder of the ether had begun
> To fold the heavens up and reinduce
> That ancient *hubris* in the dreams of men,
> Which would have slain the cattle of the sun,
> And filched the lightnings from the fist of Zeus.[12]

In one corner, then, the ship, a sort of twentieth-century Frankenstein creation—notice it has a heart and lungs—composed of equal parts technology and hubris. In the other corner, the iceberg: a little bit of that cold white mystic North

[12] E. J. Pratt, *Complete Poems*, part 1, ed. Sandra Djwa and R. G. Moyles (Toronto, 1989), 303–4.

that will claim you for its own. Here is Pratt's description, following from the birth of the iceberg in Greenland and its journey south to Labrador:

> Pressure and glacial time had stratified
> The berg to the consistency of flint,
> And kept inviolate, through clash of tide
> And gale, facade and columns with their hint
> Of inward altars and of steepled bells . . .[13]

Pratt of course had no way of knowing what the real iceberg actually looked like, so these religious images are purely fictional. Their purpose is to create an ironic illusion—the berg is not pure and holy and Christian, though it may appear so—and they are quickly melted when the berg gets far enough south:

> The sun which left its crystal peaks aflame
> In the sub-arctic noons, began to fret
> The arches, flute the spires and deform
> The features, till the batteries of storm,
> Playing above the slow-eroding base,
> Demolished the last temple touch of grace.[14]

What's underneath? Those who have followed the Nature-as-metaphor battle that raged throughout the nineteenth century—in which Wordsworthian good-mother imagery wrestled with Darwinian bad-mother imagery, and, by and large, lost—and especially those steeped in, say, Rider Haggard's *She* and Bram Stoker's account of Lucy's metamorphosis from bride to vampire, in *Dracula*, will know better than to trust anything wrapped in white. Here's what Pratt does:

> Another month, and nothing but the brute
> And paleolithic outline of a face
> Fronted the transatlantic shipping route.

[13] E. J. Pratt, *Complete Poems*, part 1, ed. Sandra Djwa and R. G. Moyles (Toronto, 1989), 305.

[14] Ibid.

A sloping spur that tapered to a claw
And lying twenty feet below and made
It lurch and shamble like a plantigrade;
But with an impulse governed by the raw
Mechanics of its birth, it drifted where
Ambushed, fog-grey, it stumbled on its lair . . .[15]

The iceberg, of course, is not alive, not really. But Pratt renders it as semi-alive, a sort of *Night of the Living Dead* zombie, complete with a face, a claw, a lair, and an 'impulse'. Pratt tried to avoid Hardyesque predestination, but, as his biographer David Pitt says, 'the more facts he unearthed . . . the more "incredible," "bizarre," "grotesque," "ironic," and "macabre" did they appear'. Pratt himself commented that it was 'as if some power with intelligence and resource had organized and directed a conspiracy'.[16] These are the intuitions he seems to have incorporated into the shambling, semi-alive, paleolithic figure of the iceberg itself—that chunk of the malignant North. Pratt doesn't say what gender it is, but we get a hint after the ship has gone down, with 1,400 people still aboard it and the band still playing. Here are the last lines of the poem:

And out there in the starlight, with no trace
Upon it of its deed but the last wave
From the *Titanic* fretting at its base,
Silent, composed, ringed by its icy broods,
They grey shape with the paleolithic face
Was still the master of the longitudes.[17]

That phrase 'ringed by its icy broods', so closely connected with hens and chicks, pushes the berg in the direction of the female. It might be argued that the word 'master' pushes it back towards the neuter or even the male; but surely five words and an end-rhyme count for more, and, anyway, the full weight of the Service tradition hangs in the balance.

[15] Ibid. [16] Pitt, *E. J. Pratt*, 148. [17] Pratt, *Complete Poems*, part 1, 338.

The first full-fledged literary—as opposed to historical—treatment of the Franklin expedition proper that I know about is an astonishing verse drama called *Terror and Erebus*, written by the poet Gwendolyn MacEwen, then in her early twenties, and initially broadcast on the CBC around 1963. Complete with electronic music and icy sound-effects, it was a hair-raiser.

Gwendolyn MacEwen had not read *Frozen in Time*, because it hadn't yet been written, and the lead-poisoning theory was unknown. But had she read Robert Service? Had she read Pratt's *Titanic*? The only extensive study of her work that I am aware of does not say. But everybody of our generation knew some Service, and Pratt was taught in high school, and MacEwen was an avid reader. However that may be, she power-packs her play with a good deal of imagery that you will not, by now, find totally unfamiliar.

The speakers in the play are Franklin himself, Crozier, the explorer Rasmussen—speaking almost 100 years after the Franklin expedition itself, and after the North-west Passage had indeed been discovered—and an Inuit character called Qaqortingneq. The story-line is book-ended by Rasmussen's commentary, and also by a sort of addendum spoken by Qaqortingneq; but the main narration is by Franklin and Crozier, and follows the historic expedition fairly closely: the three years in the ice, the sickness among the men, the death of Franklin, the abandonment of the ships and the attempt at an overland march to the south.

The language of *Terror and Erebus* is organized around several groups of metaphors, all having to do with the effect of the North itself upon the human body and imagination. One group uses images of religion—remember Service's 'vast and God-like spaces' and Pratt's altars, spires, steeples, and bells—and is introduced by Rasmussen, who says:

So I've followed you here
Like a dozen others, looking for relics
    of your ships, your men.
Here to this awful monastery
    where you, where Crozier died,
    and all the men with you died,
Seeking a passage from imagination to
    reality,
Seeking a passage from land to land
    by sea.

Now in the arctic night
I can almost suppose you did not die,
But are somewhere walking between
The icons of ice, pensively
    like a priest,
Wrapped in the cold holiness of snow . . .[18]

Here we have 'relics', 'awful monastery', 'icons', 'priest', and the 'cold holiness of snow', all in a mere dozen lines or so. There's a question which isn't asked, but is implicit—given these metaphors, what god presides here, or is being worshipped?

The second group of metaphors has to do with madness. When Rasmussen first mentions finding the bodies and bones of the lost men, he compares them to 'shattered compasses, like sciences | Gone mad'. Franklin himself speaks of the strait where they are trapped as a 'white asylum', and continues:

The ice clamps and will not open,
For a year it has not opened
Though we bash against it
Like lunatics at padded walls.[19]

If the men on the overland march stop, they will 'stand like catatonics | In this static house of bone'. Part of this imagery

[18] Gwendolyn MacEwen, *Afterworlds* (Toronto, 1987), 42–3.

[19] Ibid. 43.

of insanity has to do with the collapse of science in circumstances in which rationality and objectivity cease to have meaning because they have become useless. As Rasmussen says, 'The ice | Is its own argument,' and the men leave all their measuring implements strewn behind them: 'compasses, tins, tools, all of them. | We came to the end of science.' But part of it is traditional: going mad is what you do in the North.

The third group of images is sexual, but sexual in a very sinister way: the ships are wombs, leaving them is a forced birth, the 'giant virginal strait of Victoria' is depicted as holding the men 'Crushed forever in her stubborn loins, | her horrible house, | her white asylum in an ugly marriage'. This last passage is the only personification of geography in the entire poem, and it's noteworthy that the figure conjured up is giant, female, icy, connected with madness, and destructive: a sort of Nature white in tooth and claw. As they die, suffering from famine, insanity, exposure, and snow-blindness (they don't even have the sense to adapt Eskimo-style wooden snow-goggles), the men pray to the standard Christian God, but their prayers are prayers of despair; the implication is that they're talking to the wrong God here.

When Qaqortingneq speaks, however, there is none of this sort of language: no madness, no religious metaphors, no sinister sexuality. Qaqortingneq's speech is a simple account of how his ancestors, while out hunting seals, found the ice-locked ships, went inside them, discovered some very dead men in there, and bored holes in the sides of the ships to let the light in, causing the ships to sink. (This last detail is fairly implausible—Inuit people are not so dumb about boats, and would know that if you make a hole in one it is likely to have adverse effects upon the boat. Also, the ships were copper-hulled; boring holes in them would not have been work for an idle afternoon. But MacEwen obviously felt a poetic need for this passage: the ships of death with their cargoes of corpses—placed there by MacEwen herself, not by the historical record—settling down under the icy water. My own theory is that these ships sink in MacEwen's poem because the *Titanic* sank in Pratt's, and she just had to get those ships down to the bottom of the sea somehow, because it was so obviously—poetically—the right place for them. But I have no way of proving it.) None of the evil-Nature horrors that assail the Franklin expedition seem to bother the Inuit at all: not the ice, not the snow-blindness, not the disorientation and madness, not the lack of food. Qaqortingneq doesn't even mention the cold. MacEwen knows her ballad lore: in her drama, as in 'Lord Franklin', it's 'the Eskimo in his skin canoe' who is 'the only one who ever got through'.

It's Rasmussen who has the last word, however. At the beginning of the drama, he says to Franklin:

> The earth insists
> There is but one geography, but then
> There is another still—
> The complex, crushed geography of men.
>
> You carried all maps within you;
> Land masses moved in relation to you—
> As though you created the Passage
> By *willing* it to be.
> Ah Franklin!
> To follow you one does not need geography.
>
> . . . .
>
> The eye *creates* the horizon,
> The ear *invents* the wind,
> The hand reaching out from a parka sleeve
> By touch demands that the touched thing
> be.[20]

And at the end, Rasmussen returns to this point of view:

[20] Gwendolyn MacEwen, *Afterworlds* (Toronto, 1987), 42.

Now the great passage is open,
The one you dreamed of, Franklin,
And great white ships plough through it
Over and over again,
Packed with cargo and carefree men.
It is as though no one had to prove it
because the passage was always there.
Or . . . is it that the way was *invented*,
Franklin?
That you cracked the passage open
With the forces of sheer certainty?[21]

Here is the 'heroic' version of Franklin, with a vengeance—Franklin blasting his way through geology by thought-rays alone. This point of view—that the imagination and the will invent reality—has a corollary: if it was a success of the imagination that 'created' the Passage, it was a failure of the imagination that created the failure of the expedition itself. In simple terms: if you don't think right about the North, the ice-goddess will get you.

After MacEwen's verse play, something happened in the real world that tended to undercut the heroic version of Franklin: air travel became common, and a lot of the flights went right over the North-west Passage, where so many had died—viewed from an airplane vantage-point—like ants. My next glimpse of Franklin is in a short 1967 poem by Al Purdy, spoken from inside a jumbo jet, and called—no surprise—'The North West Passage'. It ends this way:

The North West Passage is found
And poor old Lady Franklin well
she doesn't answer the phone
tho once she traded her tears for ships
to scour the Arctic seas for her husband
but the *Terror* and *Erebus* sank long ago

[21] Gwendolyn MacEwen, *Afterworlds* (Toronto, 1987), 56–7.

and it's still half an hour before dinner
and there isn't much to do but write letters
and I can't think of anything more to say
about the North West Passage
but I'll think of something
maybe
a break-thru
to strawberries and ice cream for dinner[22]

After such a diminuendo, what resurrection is possible? Quite a few, as it happens. You can't keep a good myth down, and Franklin keeps popping up in unexpected places. For instance, the original ballad makes an appearance in Graeme Gibson's 1982 novel *Perpetual Motion*, which is set in nineteenth-century Ontario. The protagonist of this novel is an obsessive Scot named Robert Fraser, devoted to technological progress, which in him takes the form of a desire to build a perpetual-motion machine; so devoted is he, in fact, that he is willing to destroy Nature itself in order to do it. But it's not Robert Fraser who sings the Franklin ballad: it's his son Angus, who rebels against his father and, shortly after singing the Franklin song, and sickened by the slaughter of the Passenger Pigeons from which is father is making the money to finance his infernal machine, runs away and turns into a crazed wild man in the forest; becoming 'lost' himself, and taking, as it were, the side of Nature in the conflict between Nature and technology. Or being taken by Nature, claimed as one of her own. Here's the passage:

Angus interrupts with a drawn-out note, a phrase that becomes the song, a lament for the Erebus and the Terror, for Lord Franklin, beloved Franklin in the Frozen Ocean, for gallant seamen drifting like flowers beneath mountains of ice.[23]

[22] Al Purdy, *The Collected Poems of Al Purdy*, ed. Russell Brown (Toronto, 1986), 80, 81.

[23] Graeme Gibson, *Perpetual Motion* (Toronto, 1982), 213.

Again, as in MacEwen, there's a curious insistence on placing the lost Franklin expedition *down* or *under*. *Drifting* would certainly suggest that the sailors are drowned, are underneath the arctic ice-floes; yet, from all the evidence, they died on land. And *flowers* suggests something still more or less alive, or at least fresh-frozen. Gibson does not say *dead* flowers. I suspect that, again, the Franklin story and the *Titanic* story have interpenetrated one another. It was the *Titanic* that went down with a cargo of the living, not the *Terror* and the *Erebus*.

Or perhaps this dislocation has something to do with the tendency of Canadian literature to locate the oracular voice, not in a cave or on top of a mountain or in a church or temple, but under the water—or, to cite the title of a book by poet Alden Nowlan, *Under the Ice*. Did English-Canadian literature inherit this tendency from Tennyson's *In Memoriam*, haunted by drowned Arthur Henry Hallam? Or does it have a real-life basis—does it have to do with the amount of water in and around Canada, and the number of people who seem incapable of keeping their heads above it? Or have the *Titanic*, the Franklin expedition, and the Victorian predilection for drowned poets all made their contribution? I don't know. I do know that I have always been baffled by Duncan Campbell Scott's 'The Piper of Arll',[24] in which a whole crew that appears to have something to do with Art sinks to the bottom of the sea, without even any holes being bored in the ship; and I have been equally intrigued by the end of A. M. Klein's long poem 'Portrait of the Poet as Landscape', in which the prophetic poet 'shines | like phosphorus. At the bottom of the sea'. Why there? Because, apparently, that's where Klein felt it appropriate to place him.[25]

[24] Duncan Campbell Scott, in A. J. Smith (ed.), *The Book of Canadian Poetry: A Critical and Historical Anthology*, 3rd edn. (Toronto, 1957), 208.

[25] A. M. Klein, *The Rocking Chair and Other Poems* (Toronto, 1948), 50.

The very latest writer to have a go at Franklin is Mordecai Richler, in his 1989 novel *Solomon Gursky Was Here. Solomon Gursky* is a big, sprawling book that is mostly, ostensibly, sort of, about the unconventional behaviour of a very rich Jewish family of ex-bootleggers in and around Montreal; but it is about many other things as well. Richler has always been a parodist as well as a satirist, and one of the many things *Solomon Gursky* delivers is an outrageous burlesque of the Franklin story.

Richler has done his homework. Among the sources he lists at the back of the book are, wouldn't you know, *Frozen in Time*, by Owen Beattie and John Geiger. He has also delved into the files of the *Beaver*, which, it should be explained, is the venerable publication on the North put out by the Hudson's Bay Company. And, yes, he knows his ballads. There it is, at the beginning of chapter 6: the last verse of the Franklin ballad, quoted from the one you've heard this evening.

After this tip of the hat to tradition and pathos, Richler goes to town. Drawing heavily on *Frozen in Time*, he recounts the provisioning of the expedition—'thousands of cans filled with preserved meat', and so forth—but he smuggles in some foodstuffs—and a couple of extra seamen—of his own:

> And then on the dark night before they sailed out of Stromness Harbour, in the Orkneys, their last home port, there was the curious case of the assistant surgeon of the Erebus boarding with a cabin boy wearing a silk top hat, the two of them lugging sacks of personal provisions. Six coils of stuffed derma, four dozen kosher salamis, a keg of schmaltz herring and uncounted jars of chicken fat . . .[26]

The expedition duly vanishes, and Richler then recounts the search for it—the efforts of Rae and McClintock—and the finding of the longboat with its corpses and its useless provisions scattered around it. But he appends, slyly, a jar of chicken fat. And to the souvenirs collected and brought back

[26] Mordecai Richler, *Solomon Gursky Was Here* (Markham, Ont., 1989), 46.

to England, he adds 'a black satin skullcap with curious symbols embroidered on it'. Some of the stupider English—the English are quite stupid in this book—ascribe the problematical hat to an Eskimo shaman. One of Richler's twentieth-century Jewish characters proposes to his historical society that possibly this yarmulka—for that is what it is—is proof that several Jews embarked with the Franklin expedition, but he is pooh-poohed; upon which he announces that, according to 'understandably unpublished accounts', the items found abandoned on Beechey Island included 'a filigreed black suspender belt, several pairs of frothy garters, some silk panties, three corsets, two female wigs, and four diaphanous petticoats'. Funny, though anachronistic. But the historical association does not laugh, and there are fisticuffs over whether or not the Franklin expedition included Jews.

Richler settles the matter by creating a team of archaeologists, who dig up one of the buried seamen, à la *Frozen in Time.* This man, supposedly the ship's doctor, turns out to have been buried not only in a prayer shawl, but with a bunch of spurious recommendations and a phoney medical degree.

This is the *dead* Jew. The other one—the cabin-boy—gets away, and turns out to be Ephraim Gursky, grandfather of the Solomon of the title, and embodiment of every available Robert W. Service man-of-the-North cliché-image, plus a few outrages Richler has added on.

Solomon Gursky is Richler's most 'Canadian' book. In his own left-handed way, it's a reclamation of his native territory. In many of his other books, the movement is in the opposite direction: his protagonists can hardly wait to get *out* of Canada. But Ephraim is willing to commit physical injury in order to get *in*, and getting in means getting into the mythology. 'Solomon Gursky Was Here' is like 'Kilroy Was Here': Kilroy never existed, but, during the war, he was here, there, and everywhere; and so it is with the Gurskys. If there has no hitherto been room for them in such solemn tales as the Franklin expedition, Richler will jolly well make room for them, even if he has to do it by travesty. *Especially* if he has to do it by travesty, because travesty is his preferred mode. The Richler version of Franklin is amusing enough by itself, but when read in context—*against* the reverential Canadian literary treatments of Franklin—it's like a fart in church: hilarious but, well, sacrilegious. As Richler intends.

Several other novelists have written about Franklin, most notably Rudy Wiebe.[27] But now I'd like to backtrack a little, and end with another song: this time not a traditional folk-song, but a song written by the well-known Canadian singer Stan Rogers. It's called 'The Northwest Passage', it appears on a 1981 Rogers tape of the same name, and it's worth noting that Stan Rogers died in pure *Titanic* style, in an airplane fire during which he stood aside and helped others to escape. His song is a reprise of the spirit of earlier exploration, seen in its heroic aspect; the other names mentioned are all those of northern explores. Like MacEwen's Rasmussen, he speaks from the vantage-point of one for whom exploration has become a metaphor for a spiritual journey, because the real, physical puzzle has been solved and the way made easy; but it's been solved by the dead explorers, who are somehow there, incarnate, in the routes they helped to trace. Watch the chorus, in which 'the hand of Franklin' is still active, still 'reaching', still 'tracing'. Make no mistake about it, ladies and gentlemen: Franklin lives.

NORTHWEST PASSAGE

Chorus

Ah, for just one time, I would take the Northwest Passage
To find the hand of Franklin reaching for the Beaufort Sea
Tracing one warm line through a land so wide and savage
And make a Northwest Passage to the sea.

[27] Rudy Wiebe, *Playing Dead: A Contemplation Concerning the Arctic* (Edmonton, 1989)—a collection of essays—and *A Discovery of Strangers* (Toronto, 1994)—a book about Franklin but an earlier expedition.

Verse 1

Westward from the Davis Strait, 'tis there 'twas said to lie
The sea-route to the Orient for which so many died
Seeking gold and glory, leaving weathered broken bones
And a long-forgotten lonely cairn of stones.

Verse 2

Three centuries thereafter, I take passage overland
In the footsteps of brave Kelso, where his 'sea of flowers' began
Watching cities rise before me, then behind me sink again
This tardiest explorer, driving hard across the plain.

Verse 3

And through the night, behind the wheel, the mileage clicking West
I think upon Mackenzie, David Thompson and the rest
Who cracked the mountain ramparts, and did show a path for me
To race the roaring Fraser to the sea.

Verse 4

How then am I so different from the first men through this way?
Like them I left a settled life, I threw it all away
To seek a Northwest Passage at the call of many men
To find there but the road back home again.[28]

[28] Stan Rogers, from Chris Gudgeon, *An Unfinished Conversation: The Life and Music of Stan Rogers* (Toronto, 1993), 182, 183.

2
The Grey Owl Syndrome

My first lecture took as its point of departure the doomed Franklin expedition, and revolved around notions of the Canadian North as active, female, and sinister. This lecture is called 'The Grey Owl Syndrome', after that enigmatic personage Archie Belaney of Hastings, England, who emigrated to Canada, adopted the Ojibway Indians and was adopted by them, changed his name and his history, and emerged years later as Grey Owl, a world-famous naturalist, writer, and lecturer, accepted and beloved by all as what he purported to be.

I do not wish to speak primarily about the many Native characters who have appeared in the stories, poems, and plays produced by non-Native Canadians, nor about the energetic literature now being written by Native writers themselves. Instead, I will attempt to deal with that curious phenomenon, the desire among non-Natives to turn themselves into Natives; a desire that becomes entwined with a version of wilderness itself, not as a demonic ice-goddess who will claim you for her own, but as the repository of salvation and new life.

The whole subject is, of course, a minefield, especially at the moment. For the past couple of years, something called 'the appropriation debate' has been raging in Canada; and although other groups have participated in it, it has swirled most vigorously and to greatest effect around the issue of Native voice.

In its most extreme form, the anti-appropriationists have argued that non-Native writers have no right to write as if they were Natives, or even to write about Native issues, or even to put Native characters into their books (which would, of course, render Native people invisible or non-existent in the work of non-Natives). Natives are tired of being defined and spoken for by non-Natives: only Natives should be able to speak about and characterize themselves. However, many Native writers, such as Thomas King and Tomson Highway, don't endorse this viewpoint, since the argument is reversible and they want to feel free to include non-Natives in their own work.

In a more moderate form, as exemplified by, for instance, Okenagen poet and novelist Jeanette Armstrong, the argument is that non-Natives—whether writers or anthropologists—should not retell Native myths and legends without understanding them. This argument is not based on genetic or racial entitlement but on knowledge and accuracy. Underlying this point of view is a fundamental difference between white and Native notions of property: for whites, physical objects such as houses and land and money are owned individually, whereas imaginative or spiritual ones such as mythical stories are not. But for traditional Native peoples, land is controlled tribally or communally and property such as food is shared according to need, whereas certain sacred stories are controlled or 'owned' by individuals through their families; these people must guard the story, telling it only to those who are entitled by custom to hear it. Such differences in the notions of 'ownership' have caused much misunderstanding, with non-Natives raising cries of Censorship and Natives accusing them of stealing.

But what has been raised by the debate is yet another debate: that is, who qualifies as a Native?[1] After four centuries of intermingling—and there were never rigid and enforced taboos about racial intermarriage in Canada, as there were, for instance, in the American south—many Natives are more white, genetically, than they are Native; a whole people has arisen—the French-speaking Métis of the west—with a shared white–Native ancestry and culture; and many white Canadians claim, as a matter of pride, some 'Indian blood', perhaps to convince themselves that the land they live in is one they 'ought' to be living in. One might try to decide these matters culturally—that is, in order to be Indian you must live the culture—but even this becomes a problem when the disappearance of Native languages—fifty-two, but dwindling—and the advent of television, snowmobiles, cars, and planes are thrown into the equation. How 'authentic' is the culture, any more? How 'traditional'? Even if all Indians are Indian, are some Indians more Indian than others? (These question will sound familiar to the Welsh, who have for years been trying to agree on criteria for Welshness.)

To drop the unfortunate Grey Owl into the midst of this whirlpool is perhaps to subject him to even more vilification than be received when it was discovered, after his death, that he was not Indian after all. On the other hand, he and his avatars have suddenly become a lot more interesting. Why did they do it? Imitation, after all, is the sincerest form of flattery. What was it that the Indians had—or were thought to have—that the imitators wanted?

While I'm at it, I'd better seize the nettle 'nomenclature' by its thorny prickles. For a while, 'Eskimo' was out and 'Inuit' was in, but then it was pointed out that 'Inuit' referred, properly, only to people of the Eastern Arctic; so 'Inuit' has stayed

[1] For a more comprehensive discussion of this issue, please see Thomas King's intro. to *All My Relations: An Anthology of Contemporary Canadian Native Fiction* (Toronto: MacClelland & Stewart, 1990), p. ix.

for *them*, while 'Eskimo' has come back for the peoples of both Eastern and Western Arctic. 'Indian' is roundly hated by some, who prefer 'Native peoples'. Other Native writers continue to use 'Indian', although many refer to themselves as they would have before the advent of Europeans, by tribal affiliation, as 'Ojibway', 'Mohawk', and so forth. When a name is needed for everyone south of the tree-line, 'First Nations' is often used—which, however, excludes the Eskimoes. The francophones dispose of the problem with the single word 'autocthones'. I myself have given 'Native people' about equal air-time with 'Indians', although I've tended to use 'Indians' when the authors I'm dealing with do so.

Now to the tangled question of cross-dressing, and its mixed motives. I suppose there are some analogues in other colonial or post-colonial countries—white Australians who wanted to become Aboriginals, or New Zealanders who claimed kinship with Maoris, or—especially—Englishmen who wished to become Arabs—but the infatuation, on the part of Europeans and those of European ancestry, with Native North Americans and with the Indians in particular has been enormous, and of long duration. To account for it, one could mention the relative ease with which a white person *could* 'go Indian'—if you already had dark hair, an aquiline nose, and a tan, like Grey Owl, you could pass without intense scrutiny. It was a lot easier than trying to 'go African' or 'go Chinese'.

Then there was the—again relative—respect in which Native peoples have been held. The northern North American Indians—as opposed to the Aztecs and Mayans—were never enslaved, for instance. In Canada, most tribes were never exactly conquered; more typically, they were done out of their land through trickery involving legalistic paper sleight-of-hand or through simple encroachment. Apart from that, the Indians were at first an essential part of European foreign policy. Many Indian nations joined in the European wars held on North American soil, on either the French or the English side, and their warriors were highly regarded. Indian Natives conducted diplomatic missions and negotiated treaties, and alliances with them were vigorously sought. In addition, the Indians became identified with Rousseau's 'noble savage' concept, and their reputation benefited from Romanticism generally, with its love of nature and its yen for the 'primitive'. There was, always, an opposing view, which can be found in Frances Parkman's squalid Hobbesian version of Native life or in Mark Twain's sinister Injun Joe or in the scalping Red Devils so often found in early twentieth-century Western adventure stories. These two renditions of Native peoples, as either better than whites or worse—with whites being the norm, the standard for comparison—ought to sound very familiar to women, polarized as they have been until so recently into angel-wives or demon-whores. The Other is frequently a dumping-ground for anxieties, or a way of unloading our moral responsibilities by defining other people as, by nature, better behaved than we are.

The Good Indian and the Bad Indian can be seen fighting it out in the work of, for instance, Fenimore Cooper; but, on balance, the Romantic version gained ground as the actual Indians lost it, at least in the minds of safely urbanized Europeans, at least in literature. The very successful American film *Dances With Wolves*—which was preceded by a Canadian one, *Where the Spirit Lives*—has not given us a new revelation but a very old one, newly refurbished.

There is, however, another, thoroughly practical reason for this relative respect, especially in Canada. Unlike, say, the British during their invasion of India, the first Europeans in Canada literally could not have survived without the help of the Native peoples, which was frequently offered out of a spirit of generosity and honour and at considerable cost to the Natives themselves. Before his last, fatal expedition, Franklin made an overland excursion into the North-west Territories interior, on which he took a route strongly advised against by

Natives of the region. 'However, if after all I have said, you are determined to go, some of my young men shall join the party, because it shall not be said, that we permitted you to die alone . . .'. This is Akaitcho, to Sir John Franklin, as quoted by Franklin himself in *Narrative of a Journey to the Shores of the Polar Sea.*[2] Franklin went anyway, and was saved from starvation only by Akaitcho and his men, who fed the Franklin bunch on grease soup. Author M. T. Kelly, an expert on the literature of Canadian exploration, has this to say: 'Anyone who is familiar with them finds a pattern in the trips that is often the same. The adventurers may be known as heroes, yet they were always led, guided, literally carried and often saved from death by native people.'[3] And here is D'Arcy Arden, speaking of explorer John Hornby: 'He was a stubborn man. That's why he's dead. I said, "You go there where there are no Indians, Jack, and you'll die. Every time you've starved, Jack, an Indian has come to your assistance. You get away from those Indians and you'll die like a rat."'[4]

There is a story that parallels the Franklin expedition story, and is just as true, just as dramatic, and in many ways more representative. The remarkable thing about it is not the story itself but the fact that, to my knowledge, nobody has turned it into a poem or a novel. It's the *Polaris* story, and it goes like this:

In 1870 the *Polaris*, under the command of explorer Charles Hall, set out on an expedition to reach the North Pole. The ship wintered over in a bay which was named, by Hall, Thank God Harbour; here Hall himself died, poisoned by arsenic, which was probably administered by the ship's physician. Hall was not well liked, and his obsessive desire to reach the Pole was not shared by most of the crew members. The next spring

[2] Sir John Franklin, *Narrative of a Journey of the Shores of the Polar Sea in the Years 1819, 1820, 1821, and 1822* (1823; repr. Rutland, Vt., 1970), 225.

[3] M. T. Kelly, in conversation between M. T. Kelly and Margaret Atwood.

[4] See George Whalley, *The Legend of John Hornby* (London, 1962), 127.

the ship diddled around a little and then got caught in pack-ice, which drifted it 300 miles to the south. In October there was a severe gale which threatened to destroy the ship by pounding chunks of ice against its hull. Some of the crew abandoned it, and were swept away from it on an ice-floe. Marooned on the floe were nine whites and nine Inuit. Two of the Inuit were women, and five were children, one of them only two months old.

On a piece of drifting ice, heading into winter, their chances of survival seemed nil. They had some supplies: a couple of small boats, some guns, some ammunition, a tent, six sled-dogs, and 2,400 pounds of food, such as ship's biscuit and pemmican. Still, this was not enough to keep eighteen people alive through at least six months of arctic winter. There was also the problem of shelter.

The survival of this party was due entirely to the efforts of the Inuit men. It was they who built snow huts for shelter, and went hunting—successfully—for seals. The white sailors, on the other hand, were worse than useless. They contributed nothing to the general welfare, but played cards and slept, stole supplies, and were insubordinate and ungrateful. Tyson, the officer in charge, could not discipline them because they were too numerous and well-armed. Through all of this, the Inuit hunters continued to feed them, as well as their own families, on the seals, birds, and bear they managed to shoot. Then the ice-floe melted and the entire party had to make its way to the next one, using the one boat they had left. (The other boat had been brilliantly chopped up by the white sailors for firewood.) When the party was finally rescued, after six and a half months on the ice, not one person had perished.[5]

The heroes of this episode were the Inuit hunters; the whites made a terrible showing, which is perhaps why the *Polaris* survivors have never been celebrated, as they say, in

[5] W. Gillies Ross, *Arctic Whalers, Icy Seas: Narratives of the Davis Strait Whale Fishery* (Toronto, 1985), 196–9.

song and story. But, as I've indicated, this event was not unique. It was part of a general pattern. It's against this kind of historical reality that the concept of 'going Indian'—in a more literary sense—must be read.

Two essentially nineteenth-century writers illustrate the enormous popular appeal of this idea, and were effective precursors of Grey Owl: John Richardson, author of the romance *Wacousta*, which appeared in 1832; and Ernest Thompson Seton, who, unlike Grey Owl, did not attempt the rather modest feat of turning himself into an Indian, but pursued the greater ambition of turning everybody else into Indians, instead.

Richardson's *Wacousta* has a long and involved plot of the Walter Scott or Fenimore Cooper variety, with fainting damsels to match. It is set at the time of the eighteenth-century Pontiac wars, and revolves around Pontiac's attempt to capture Forts Detroit and Michilimackinac from the British. It was phenomenally popular, both in Canada and in the United States, until at least the mid-1930s, and was pirated, condensed, reprinted numerous times, and even turned into a Broadway play. The title character is a passionate and frightening apparition, who has disguised himself as an Indian warrior—although he is in reality a high-born Cornishman—in order to revenge himself upon the man who, many years earlier, had stolen his fiancée. This character was inspired by a real man, John Norton, from West Augusta—which, shortened, yielded the name 'Wacousta'—who had adopted Indian dress and customs out of revulsion at his treatment by white society.[6] But Wacousta the character far outdoes any possible model. In the British part of the narrative—which takes place in Scotland—he's a wronged and pathetic figure, but as Wacousta he is possessed of supernatural strength and ferocity, and is somehow, well, *taller*.

[6] David R. Beasley, *The Canadian Don Quixote: The Life and Works of Major John Richardson, Canada's First Novelist* (Erin, Ont., 1977), 11.

Becoming an Indian, in this novel, is only a means to a vengeful end—what Wacousta really wants is to polish off the three children of his rival—and doesn't have a lot to do with woodcraft or surviving in the North; but it confers the same benefits as Captain Marvel's long johns or Batman's mask—that is, extra strength and the ability to terrify your enemies. Richardson is ambivalent about the merits of Indians *vis-à-vis* whites, and about the virtues of the wilderness as opposed to the 'civilization' to be found within the walls of the British forts; and in this he was perhaps reflecting his own mixed ancestry: his grandmother was in fact an Indian, and dressing up as an Indian in order to wreak revenge on white society may have been something he himself secretly wanted to do, although, as a soldier who fought on the British side in the War of 1812, he would have felt obliged to censor the impulse. As it is, the plot comes out a draw: the Indians don't take the fort, but Wacousta bags two of the three children, plus his hated rival, before he himself is put out of his misery.

Wacousta was recently given a resurrection, in the form of a play called *Wacousta!*—that's with an exclamation mark—by the Ontario playwright and poet James Reaney. In this play, the moral balance shifts in favour of Wacousta—his rival, de Haldimar, is portrayed as even more of a dishonest, opportunistic, and oppressive cold fish than he is in the novel—but especially in favour of the Indians. The speech Pontiac is given, to explain why he wants to destroy the British forts, is much stronger in its language than anything the British have to say either against the Indians or in their own defence:

> the English red coat hunting dogs have set eggs in our hunting grounds, eggs which they call forts . . . nine serpent eggs . . . These nine eggs are filled with straight lines, freeways, heartlessness, long knives, minuets, harpsichords, hoopskirts, death, disease, right-angled extermination. One by one . . . we crushed their forts before they could sit on them long enough to hatch their murderous culture. . . . In these forts, tough hard-faced foreign devils,

worshippers of the sky-demon Jehovah, shut their gates fast hermetically sealed.[7]

Repeatedly, the geometrical straight lines and right-angles of the whites are set against the organic curves of the Indians and the forest, to the former's disadvantage. At the end of the play, with Wacousta and his rival both dead, Ellen, Wacousta's mistress—who is white—walks off into the bush to join the Indians, bearing Wacousta's child, and 'we hear the cry of a loon'. The loon cry makes the point. Ellen's last words, which she speaks twice—once to the Indian woman Oucanasta, and once to us, the audience—are, 'I am sick of hating them alone.' Effectively, although both Wacousta and his mate are white, their child will be an Indian; and we, the audience, are invited to become Indians too: when Ellen tells us, 'I am sick of hating them alone,' she's inviting us to join her on the Indian side.

A footnote to *Wacousta!* is that one of the students who helped with the play's creation and production was Tomson Highway, the Cree playwright who has since gone on to enormous success with his plays *The Rez Sisters* and *Dry Lips Oughta Move to Kapaskasing*. According to Highway, it was *Wacousta!* that gave him the idea that he, too, might be able to create plays out of his own experience and locale. So a play about a fake Indian was inspirational to a real one. Such are the ironies of literature.

In the late nineteenth and early twentieth centuries, as the age of the colonial wars receded, motifs of revenge and warfare gave way to themes of nostalgia; and as the age of the explorers receded as well, living like the Natives in order to survive in the wilderness was translated into living like the Natives in the wilderness in order to survive. Survive what? The advancing decadence, greed, and rapacious cruelty of white civilization, that's what. This is the subtext of the lives and work of both Grey Owl and Ernest Thompson Seton—or

[7] James Reaney, *Wacousta!* (Toronto, 1979), 37, 38.

Black Wolf, as he was sometimes known—and it's this subtext that has surfaced again in the Canadian environmental movement of the 1980s and 1990s—a movement that has yet to claim Seton and Grey Owl as ancestors, although it does acknowledge Farley Mowat and Fred Bodsworth.

As well as being Indian impersonators, both Grey Owl and Black Wolf were conservationists and naturalists well in advance of their time; Seton, for instance, practically invented the animal story as told from the point of view of real animals, and anticipated many a Konrad Lorenz's findings about animal behaviour. But it's his Indian interests that concern us at the moment.

Ernest Thompson Seton was born in England in 1860, and moved with his family to Canada when he was 6. He read Fenimore Cooper's *The Last of the Mohicans* when he was 12, and, although he had never seen an Indian, he 'recognized the adventure potential of the Indian as a make-believe character for himself'.[8] After writing a play for his friends in which the Indian, played by himself, was for a change victorious, he attempted to organize his friends into an Indian tribe. These early experiences were the basis both for his 1903 boys' novel *Two Little Savages*, which describes the attempts of two young boys to play seriously at being Indians and to teach themselves woodcraft, and for Seton's later formation of the Woodcraft Indian movement.

The attitude towards Indians in *Two Little Savages* is almost entirely adulatory. In the debate between Yan and Sam about the relative merits of white hunters versus Indian ones, Yan, taking the side of the Indians, comes out on top: 'A White hunter can't trail a moccasined foot across a hard granite rock. A White hunter can't go into the woods with nothing but a knife and make everything he needs. A White hunter can't hunt with bows and arrows and catch game with snares, can

[8] Betty Keller, *Black Wolf: The Life of Ernest Thompson Seton* (Vancouver, 1984), 57.

he? . . . we want to be the best kind of hunters, don't we? . . . Let's be Injuns and do everything like Injuns.'[9] It is, however, the *old-time* Indians that Seton wants to imitate—flint-arrowhead Indians, pre-gun, pre-matches Indians. When the boys come upon some modern Indians, the encounter is short and not very productive.

In his Woodcraft Indian movement—which Seton started as an experiment in boy-control with a group of youngsters who were trespassing on some land he'd bought—Seton enshrined this ultra-positive attitude towards all things Indian. The preface to the 1912 *Book of Woodcraft* is a passionate hymn of praise to the virtues of the Ideal Indian—'the highest type of primitive life . . . a master of woodcraft, and unsordid, clean, manly, self-controlled, reverent, truthful, and picturesque always'. Seton proposes to use the Indians as a 'model for white men's lives'[10] and a cure for their over-civilized ills. This sort of apologia was needed: by the end of the century, the reputation of the Indians had seriously declined, and it needed all the help Seton could give it.

The story of the Woodcraft Indians is a sad one. Seton developed the organization along democratic lines—grown men, for instance, were never chiefs, only 'medicine men', or counsellors—and the boys elected their own leaders and conferred their own honours upon one another. This movement was later stolen from him by Baden-Powell and converted into the Boy Scouts, with its hierarchy and its quasi-military organization—so useful for producing future soldiers, it was thought in 1914. Seton was told that the American public had come to dislike Indians as role-models for its kids; and was squeezed out of his advisory position, by, among others,

[9] Ernest Thompson Seton, *Two Little Savages: Being the Adventures of Two Boys who Lived as Indians and what they Learned* (Garden City, NY, 1959) (1912; repr. Berkeley, 1988), 83, 84.

[10] Ernest Thompson Seton, *The Book of Woodcraft* (1912; repr. Berkeley, 1988), 57.

Theodore Roosevelt, who denounced his erstwhile friend as a pacifist.

But despite his defeat at the hands of Baden-Powell's Boy Scouts, Seton's influence was pervasive and lasted a long time. 'Indianism' of his type, in diluted form—the dressing up, the adoption of 'Indian' names, the playing of games, the singing of songs—was a feature of summer camps and of Girl Guide and Boy Scout groups in Canada well into the 1940s and 1950s. A great many children, over the space of fifty years, were invited to believe that they were metaphorically Indian, and to invest some of their own identity in this notion. 'My paddle's clean and bright,' they sang, 'flashing like silver'[11]—which would have had entirely other connotations to someone with a British public school background. Or,

> Land of the wild goose, home of the beaver,
> Where still the mighty moose wanders at will,
> Blue lake and rocky shore, I will return once more
> Boom biddy boom boom boom biddy boom boom
> boo-oo-oo—oo-oom.[12]

When such children read Pauline Johnson's 'The Song my Paddle Sings'[13] in their school readers—as they did, until the 1960s—they knew that the paddle in the song was not just Pauline's paddle but their paddle too.

The irony of this general movement is that it was taking place during a period of intense but concealed guerrilla warfare conducted against Indians—when the North was being increasingly invaded by the forces of commercialism, when treaties were being violated, lands appropriated, rights denied. It was the same era in which Indian children were being virtually kidnapped in the interests of 'assimilation', hauled off to residential schools far from their own families, told that their

[11] Traditional song sung at summer camp. [12] Ibid.

[13] E. Pauline Johnson, *Flint and Feather: The Complete Poems of E. Pauline Johnson (Tekahionswake)* (Toronto, 1969), 31.

customs were evil, beaten up for speaking their own languages, and subjected—it has now been revealed—to all sorts of child abuse. This kind of cultural suppression was not unique to Canada, but it was horrible all the same. However, rather than condemning Seton's Indian-worship as sentimental and hollow make-believe, the contrast reinforces its value; at least Seton created a generation of quasi-Native children, some of whom would grow up to be ardent conservationists and sympathetic to Indian claims once they had made their way on to the agenda again. It's harder to collude in the destruction of a people and a way of thinking if you feel that this people and this way of thinking are partly your own.

Seton knew where the boundary was between his created 'Indian' world and that other world in which he was a rich landowner, a friend of the famous, and a noted author and naturalist. He knew that he was not really 'Black Wolf'. He knew that much of what he was doing in his 'Indian' world was both advocacy and play, and that the sort of Indian he wanted to emulate existed more in the past than in the present. Grey Owl, on the other hand, was bent on the obliteration of boundaries. He did not want to have two identities, only one: the Indian one. He was willing to go to great lengths to acquire it. And when Grey Owl came into being, Archie Belaney, in his own mind, virtually ceased to exist.

Archie Belaney was born in England in 1888, the son of a scapegrace father, who shortly disappeared to America, and a somewhat dubious mother; but his father was from a respectable family, and Archie's two aunts wrested the child away from his parents and brought him up, 'by hand', as Dickens would say. Like Seton, Archie was influenced by Fenimore Cooper, and like Seton he played Indian games as a boy; but his fantasies were of an even more intense order. He was determined, not just to play at being an Indian, but to become one, and he ran away to Canada while still a teenager in order to do just that. He had hardly any money, but he worked his way north. After a classic episode in which he almost froze to death and was rescued by 'real' Indians, he set himself to learn Ojibway, dress appropriately, and scrape his living by guiding and trapping, just like the Indians he'd met. He put it about that he was the son of a Scottish father and an Apache mother, and he was believed. Eventually he was adopted into a local tribe and given a new name, 'He-Who-Travels-By-Night', or Grey Owl.

The early twentieth-century Canadian North that Grey Owl entered in the years just before and just after the war was dual in nature. On the one hand, the 'virgin' wilderness was giving way to settlement, was being 'opened up'. (Consider that phrase in relation to the North as female and see what you get.) On the other hand, the North had become a fetish, among those who no longer lived in it but were eager to prove their authenticity and manhood by stepping back into it. I say 'back', because to enter the wilderness is to go backwards in time; which may account for the relentlessly elegiac and archaeological streak in Canadian literature.

This Northern fetishism was so well entrenched by 1910 that it's satirized in 'Back to the Bush', one of the pieces in Stephen Leacock's *Literary Lapses.* Billy, Leacock's friend, has 'the Mania of the Open Woods', and is trying to drag Leacock with him.

> 'How do we go, Billy, in a motor-car or by train?'
> 'No, we paddle.'
> 'And is it up-stream all the way?'
> 'Oh, yes,' Billy said enthusiastically.
> 'And how many days do we paddle all day to get up?'
> 'Six.' . . .
> 'Glorious! and are there portages?'
> 'Lots of them.'
> 'And at each of these do I carry two hundred pounds of stuff up a hill on my back?'
> 'Yes.'

'And will there be a guide, a genuine, dirty-looking Indian guide?'
'Yes.'
'And can I sleep next to him?'
'Oh, yes, if you want to.'
'And when we get to the top, what is there?'
'Well, we go over the height of land.'[14]

Grey Owl worked for years as one of these 'genuine, dirty-looking' Indian guides, carting white stockbrokers around in the bush. After various adventures and brushes with the law, he married a 'real' Indian, Anahareo; but she was less interested in being 'real' than he was. As Grey Owl tells it in *Pilgrims of the Wild*, she wasn't so keen on going off for months into the wilderness, and she wanted Grey Owl to update his buckskin-and-fringes wardrobe and get a radio. Finally, she objected to the slaughter involved in trapping, and it was her squeamishness that drove Grey Owl into the life of a naturalist-writer—he had to do something else to make money. Through the couple's efforts to protect the almost-trapped-out Canadian beaver, Grey Owl was 'discovered' and became famous as a lecturer just before the Second World War. His biographer, Lovat Dickson, explains his appeal: 'In contrast with Hitler's screaming, ranting voice, and the remorseless clang of modern technology, Grey Owl's words evoked an unforgettable charm, lighting in our minds the vision of a cool, quiet place, where men and animals lived in love and trust together.'[15]

Grey Owl, like Dylan Thomas, was exhausted by the strain of his lecture tours. He died in 1938, at the age of 50, and within a couple of days he was being vilified as a hoax, fraud, and impostor, because it had been discovered that he wasn't a real Indian. His motives were disregarded and the important work he'd done as a conservationist was brushed aside by a

[14] Stephen Leacock, *Literary Lapses* (Toronto, 1971), 120.
[15] Lovat Dickson, *Wilderness Man: The Strange Story of Grey Owl* (Toronto, 1973), 5.

public that felt it had been cheated out of its pet talking Indian. The spectre of authenticity had been raised, and it overshadowed everything else.

The continuing saga of Grey Owl is to be found in the literary afterlife he, like Franklin, has led in the work of other authors. His initial sad fall from popular grace is reflected in Alice Munro's 1968 story 'The Dance of the Happy Shades', in which he turns up at an awful recital thrown by an old-fashioned Toronto piano teacher. After the children have played their selections, the teacher hands out gifts to her pupils:

> where did she find such books? They were of the vintage found in old Sunday-school libraries, in attics and the basements of second-hand stores, but they were all stiff-backed, unread, brand new. *Northern Lakes and Rivers, Knowing the Birds, More Tales by Grey-Owl, Little Mission Friends.*[16]

By the mid-1960s Grey Owl was no longer even a scandal, but had dwindled into an outmoded prop of shabby gentility.

However, someone with the eccentric vitality of Grey Owl—and, more to the point, someone who lived out such a deeply rooted collective dream—is hard to keep buried. Back he comes again, as a focal point in Robert Kroetsch's 1973 novel *Gone Indian.* This title—as is the habit of respectable titles in the late twentieth century—has at least five meanings. First, the central character tries to 'go Indian', that is, become an Indian; second, the Indians—in their old, romantic phase—are 'gone'; third, the protagonist himself not only tries to turn himself into an Indian, but is 'gone', that is, crazy. (In the slang of the time, this could be a compliment.) Fourth, at the beginning of the book he is 'gone', as in colloquial hippie talk 'I'm gone, man'—that is, he has left his accustomed locale and taken off for where he hopes the wilderness will be; and at the

[16] Alice Munro, *Dance of the Happy Shades* (Toronto, 1968), 215, 216.

end of the book, he is truly gone, because he has vanished completely.

The unlikely hero of this burlesque performance—which is what used to be called, on the jacket, a 'wild extravaganza' or perhaps a 'darkly surrealistic comedy'—is named Jeremy Sadness. Jeremy is after Jeremy Bentham, that embalmed and upright Utilitarian,[17] and Sadness is self-explanatory. Jeremy Sadness is a graduate student in English. For some reason, male graduate students in English are usually portrayed, these days, as weak, twisted, angry, and dreaming fools, whereas when they obtain teaching positions they metamorphose into power-hungry and sadistic satyrs. Female graduate students in English, on the other hand, are not weak but strong, in a metallic way, and are apt to be shown as humourless ideologues, unless they are brought to see the errors of their ways through the love of a good man, like Robertson Davies's Maria in *The Rebel Angels*. In any case, Jeremy Sadness runs true to form. His sadness comes from his discomfort with who he is: he doesn't want to be a graduate student in English. Instead, he wants to be Grey Owl. Dressed for the part of his Grey Owl braids, he announces this to the Customs Official at the Canadian–US border; because, in addition to his other handicaps, poor Sadness is a Yank, and a painfully honest one at that. 'Purpose of trip?' says the Customs Official.

> 'I want to be Grey Owl.'
> 'I beg your pardon, sir?'
> 'Grey Owl. I want to become—'[18]

Here the narrative is taken over by Jeremy's thesis adviser, Professor Mark Madham, one of the aforementioned breed of power-hungry sadistic satyrs, who comments:

[17] Bentham arranged in his will to have himself stuffed after death and to be wheeled out for an annual banquet in his honour.

[18] Robert Kroetsch, *Gone Indian* (Toronto, 1973), 6.

Jeremy Sadness might have chosen no end of frontiersmen to embody his dream of westward flight. Curiously, he chose a model from the utmost cultivated shores of the civilized world. Given as he was to self-deceiving self-analysis, he believed that his life's predicament found its type in Grey Owl. He was almost anally fascinated by that quick-tempered English lad who left Victorian England, disappeared into the Canadian bush, and emerged years later as Wa-Sha-Quon-Asin.

He-Who-Travels-By-Night.

The possibility of transformation, I must recognize, played no little part in Jeremy's abiding fantasy of fulfilment. It gave him, in the face of all his inadequacies, the illusion of hope.[19]

Jeremy Sadness is supposed to be on his way to an academic job interview in Edmonton, but his suitcase gets muddled with someone else's at the border and he traces it to the northern town of Notikeewin (whose first two syllables yield 'naughty', and whose first and last give us 'no win', which pretty much sums up the plot). Grey Owl is not Sadness's only obsession; his other one is sex, specifically, sexual failure. He seems to be able to copulate only while standing up; and he has another, more standard fear:

Jeremy Sadness was a weird young tool, if ever God made one. And surely not the least of his peculiarities was a sexual anxiety that pervaded much of his behaviour; arriving in the land of his imagination, he conceived the unlikely notion that his prick might freeze off.[20]

Immediately after this scene, Jeremy comes upon a tableau, consisting of the entire Wild North-west—Indians, buffalo, dog teams, trappers—all sculpted in ice. The True North has given way to an imitation of itself, done up as display for tourists as part of the Notikeewin Winter Festival. This imitation North is what has become of Grey Owl's wilderness, and Jeremy blunders through it, from cliché to cliché, as Kroetch plays with well-established motifs, refurbishing them

[19] Ibid. 6, 7. [20] Ibid. 17.

for his own purposes. Jeremy even meets up with a woman who, as Jeremy describes her, is the North personified, à la Robert Service: 'cold, blank, oblivious, whimsical, amoral, stark'. And he finally spots a real Indian, who is eating pancakes at the First Presbyterian Church's annual Pancake Breakfast. This encounter has disappointing results:

> 'Did you ever hear of Grey Owl?' I shouted at him. . . .
> 'Did you ever run into Grey Owl?'
> The Indian looked up from his plate, looked across the table at his wife. 'Grey Owl?' he said.
> His wife giggled.
> One of the little boys saw his chance. 'Why is his hair that way?'[21]

But Jeremy finally gets his wish, or a version of it. He is mistaken for a real Indian by a group of drunken, bigoted whites, who beat the stuffing out of him and leave him to freeze in the snow. True to historical pattern, the Natives save him from a death; he is rescued by the *real* real Indian, the one who didn't recognize Grey Owl, who sympathizes with Jeremy because, although Jeremy isn't a real Indian, he's been treated as if he is one—that is, badly. This causes Jeremy to lapse into a meditation of what 'being an Indian' means, and how he came by his Grey Owl obsession, in the course of being forced into the Indian role during upsetting childhood games in New York:

> I didn't want to be the Indian at all. They told me, You be the Indian, Sadness. We'll hunt you down. No matter where you hide, we'll hunt you down. We'll kill you. And they threw broken bricks and they tied me up and stuck lit matches into the seams of my shoes. . . . So the tailor across the hall from my mother's apartment brought me in his books of Grey Owl; one by one, he brought them. . . . He gave me his dream of the European boy who became . . . pathfinder . . . borderman . . . the truest Indian of them all.[22]

[21] Robert Kroetsch, *Gone Indian* (Toronto, 1973), 65.
[22] Ibid. 94.

This is a curious series of questions. Being 'the truest Indian' obviously has two sides to it. You can be 'the truest Indian' by being the most adept, even though you are a non-Indian; that is, Jeremy's dream is capable of realization. Or you can be 'the truest Indian' by having, as the *real* real Indian says, 'the supreme piss' kicked 'right out of you'. Poor Jeremy doesn't even want to be an *Indian*—he wants to be a white man who wanted to be an Indian. In Kroetsch's funhouse hall of mirrors or parody of a Platonic universe, Jeremy aspires to be an imitation of an imitation, who may nevertheless be 'the truest'. As for the *real* real Indian, he admits that he does remember Grey Owl after all. What he remembers about him is his bad temper and violence. 'Grey Owl would be proud,' he says, after Jeremy's fight. 'He was a good fighter . . . He killed a man himself one time, in a fight. . . . He was quick with a knife, Grey Owl. He liked to drink. He liked women.' Jeremy is not sure about this violent image. 'He killed himself', he whispers. 'He killed Archie Belaney. Then he became Grey Owl.'[23]

Jeremy, however, doesn't succeed in his attempt at imitation. After falling in love with a woman called Bea, one letter short of Bear—at one point she's dressed up like this animal—and after a dream vision in which he discovers that the reason he can only copulate standing up is that he's secretly a buffalo, Jeremy heads off on a snowmobile and disappears. Carol, his wife back in New York State, has one explanation: 'knowing Jeremy's passion to become an aboriginal of some variety, she supposes he was quite recklessly and irrationally heading for the true wilderness'.[24] Professor Madham, however, postulates that Jeremy, who 'couldn't steer a toy wagon in broad daylight', got lost in a blizzard and fell off a railway bridge. Like Grey Owl, he killed his 'white' self; unlike Grey Owl, he didn't end up with the other, 'true' self he wanted. It's

[23] Ibid. 100. [24] Ibid. 153.

the word 'true' that haunts this book—as it, and its cousins 'real' and 'authentic', haunt the entire white-into-Indian enterprise. The other words that haunt both book and enterprise are 'illusion' and 'hope'.

M. T. Kelly is even harder on Grey Owl. Most of Kelly's work revolves around the Canadian North. Like Grey Owl, he did not grow up in 'the country of his imagination', and, also like Grey Owl, he fulfilled a childhood dream by remaking himself from an urbanite into an accomplished canoeist and backwoods *aficionado.* Perhaps because of these affinities, he reacts strongly against poor old He-Who-Travels-By-Night in a story from his 1991 collection *Breath Dances Between Them.* The story is called 'Case Histories', and is about a canoe trip undertaken by three old canoeing buddies and a fourth person, a naturalist named Allison. Allison has advertised in an issue of *Wildculture* magazine for some companions to go down the Mississagi River with her, 'following Grey Owl's route'. The story begins with Jones, one of the male canoeists, quoting from Grey Owl, a passage in which Grey Owl laments the passing of the old wilderness ways: 'the rollicking chorus of the canoe brigades is replaced by the pulings . . . of pale emasculates'.[25] This sets off a spate of aggressive joking in which the men accuse one another of being 'pale emasculates', and Mick, the narrator, loses his temper:

> 'I come in peace, brother,' . . . I parodied with revulsion Grey Owl's greeting to the King of England. I tried to keep in mind that Grey Owl was an early conservationist, but then I let myself go: another tall, twisted, imperial phoney, a Brit who 'seeded' native women, then abandoned the children, a violent sociopath only concerned with himself, *his* reactions to the country, *his* feelings. I saw a six-foot child in a canoe acting wilfully, turning aside from the things he'd done to other people, turning away.
>
> He lied. He lied about the river, he lied about the rapids, at least he exaggerated; he couldn't be trusted. I thought of his books, his

[25] M. T. Kelly, *Breath Dances Between Them* (Toronto, 1991), 130.

picture of an Indian camp: all pastels, twilight behind the white pines, a movie set with obvious, false, backlighting.[26]

Mick's rage is prophetic. The canoe trip becomes more and more of a disaster, as Allison controls and manipulates the three men, flirting with them and ignoring them in turn, and exercising power by lecturing them on wild flowers and treating them to breathy little sermonettes on natural beauty and the sense of wonder, in a Grey-Owlish way. In fact, Allison is a sort of grotesque female version of Mick's idea of Grey Owl—an egomaniac, uninterested in the feelings of others, a child acting willfully—and she manages to destroy both the men's sense of solidarity and any spontaneous response they might have had towards their surroundings.

Part of Mick's rage is triggered by Grey Owl's 'pale emasculates' remark, because being a 'pale emasculate' is exactly what these men fear. Their canoe-tripping has been a search for the authentic, that Romantic commodity which is forever receding out of reach. Like Seton—and like Grey Owl, who kept driving further and further into the wilderness as it vanished behind him—their desire is for a pre-formulated idea of wilderness, and their anger is occasioned by the discrepancy between that and the actuality they experience. Like Jeremy Sadness, Mick's deepest fear is of not being real; the dilemma of all those who long for authenticity by identifying with the wilderness, and with their own idea of what an Indian should be, is that they can only be real, in their own terms, by turning themselves into something they are 'really' not.

Gwendolyn MacEwen—who, you'll remember, is the author of *Terror and Erebus*, the radio drama about Franklin—was fascinated all her life with explorers who left their home bases and wrecked themselves in foreign territories, and with escape artists, shape-changers, and those who concealed their identities or went through the world in disguise. Grey Owl makes his

[26] Ibid. 131.

appearance in her last book, *Afterworlds*, where he is more gently treated than he has been by the authors previously mentioned. He was a natural for MacEwen, for whom tricksters and illusionists—those who live two realities—are not emblems of the ersatz, the phoney, but true expressions of human doubleness, of two-handedness, of a sometimes tragic but always necessary duality. For MacEwen, Grey Owl is a quester in search of himself, a doomed hero who renders himself alien both to his original homeland and to his adopted space. her poem is called, not 'Grey Owl Poem', but 'Grey Owl's Poem': that is, it is not *about* Grey Owl but, in a more friendly vein, *for* him:

GREY OWL'S POEM

There is no chart of his movement through the borrowed forest,
A place so alien that all he could do with it
was pretend it was his own
And turn himself into an Indian, savage and lean,
A hunter of the forest's excellent green secret.

For all his movements through the forest were
In search of himself, in search of Archie Belaney,
a lone predator in London
Telling the very king: *I come in peace, brother*.
(The princess thinking how alien he was, how fine.)

Stranger and stranger to return to the forest
With the beavers all laughing at him, baring
their crazy orange teeth
And the savage secret—if there ever was one—
Never revealed to him. Stranger and stranger to return to

The female forest, the fickle wind erasing his tracks,
The receding treeline, and the snowbanks moving and moving.[27]

Grey Owl's longing for unity with the land, his wish to claim it as homeland, and his desire for cultural authenticity

[27] Gwendolyn MacEwen, *Afterworlds* (Toronto, 1987), 72.

did not die with him; not at all. All three preceded him and survived him, and are still very much with us. It was one of the accepted truisms of Canadian literary criticism in the 1950s and 1960s that the Canadian poet's task was to come to terms with the ancient spirit—that is, the Native spirit—of the land whites had not yet claimed at a deep emotional level, and this is exactly what John Newlove attempts in his long 1968 narrative poem 'The Pride'. The poem begins with images of the North and the old west, continues through legends, and ends with a statement of relationship and indeed of unity:

> not this handful
> of fragments, as the indians
> are not composed of
> the romantic stories
> about them, or of the stories
> they tell only, but
> still ride the soil
> in us, dry bones a part
> of the dust in our eyes,
> needed and troubling
> in the glare, in
> our breath, in our
> ears, in our mouths,
> in our bodies entire, in our minds, until
> at last
> we become them
>
> in our desires, our desires,
> mirages, mirrors, that are theirs, hard-
> riding desires, and they
> become our true forebears, moulded
> by the same wind or rain,
> and in this land we
> are their people, come
> back to life again.[28]

[28] John Newlove, *The Fatman: Selected Poems 1962–1972* (Toronto, 1977), 74.

Well, maybe. The difficulty is—as several pointed out when this poem appeared—that you don't really need to resurrect someone who isn't dead. This kind of 'claiming kin', based though it is on desperate longing, drives some Native people wild and makes others hoot with sardonic laughter. The fact is that they may not particularly *want* to be our ancestors. Consider Buffy Ste Marie's song 'Where have the Buffalo Gone?', which lambastes the whites for simultaneously claiming kin and neglecting and impoverishing their so-called relatives. Or consider Maria Campbell, speaking in *The Book of Jessica*: 'Using the word ghost is good because that's what the old people say when they talk about white people in this country: "Ghosts trying to find their clothes."'[29]

But like it or not, the wish to be Native, at least spiritually, will probably not go away; it is too firmly ingrained in the culture, and in so far as there is such a thing as a Canadian cultural heritage, the long-standing white-into-Indian project is part of it. Perhaps the thing to do with it is not to repudiate it or ridicule it as naïve, sentimental, and embarrassing—a matter of grown men playing feather dress-ups—but to take it a step further: if white Canadians would adopt a more traditionally Native attitude towards the natural world, a less exploitative and more respectful attitude, they might be able to reverse the galloping environmental carnage of the late twentieth century and salvage for themselves some of that wilderness they keep saying they identify with and need. Perhaps we should not become less like Grey Owl and Black Wolf, but more like them.

I would like to close with an excerpt from another poem by Gwendolyn MacEwen, which sums up the dilemma—the feeling of being alien, of being shut out, and the overwhelming wish to be let in. Her poem is called 'The Names'.

[29] Linda Griffiths and Maria Gampbell, *The Book of Jessica: A Theatrical Transformation* (Toronto, 1989), 96.

THE NAMES

We want to pretend that you are our ancestors—
you who are called
Wolf In The Water, Blue Flash of Lightning, Heaven Fire,
Black Sleep—

You who have no devil, no opposite of Manitou.

You who are hiding behind your names, behind
closed doors of thunder
And will not let us in.

. . .

You who never knew the evil in us, you who have
no opposite of Manitou,
Come out from behind the thunder and embrace us—
All we long to become, all we have ever known of ourselves.

Before you are gone from our eyes forever—
(you who are certainly not our ancestors)
Teach us our names, the names of our cities.

No one ever welcomed us when we came to this land.[30]

[30] MacEwen, *Afterworlds*, 73.

3

Eyes of Blood, Heart of Ice: The Wendigo

In my first lecture, I talked about some of the patterns of belief and imagery that have accumulated around the idea of the Canadian North, and about how these patterns have recurred, with variations, in the work of a number of writers. One of the patterns had to do with going crazy in the North—or being driven crazy by the North—and now I would like to discuss one particular incarnation of this concept—the Wendigo. This creature also serves as a specific case, illustrating the extent to which Native motifs have infiltrated non-Native literature and thought.

So let us begin with the figure of the Wendigo—or is it a figure? Perhaps it is also a verb, a process; it can certainly be used as such. One can 'go Wendigo', and one can 'become Wendigo'. What then is 'Wendigo', what does it do, and what meanings accrue to it when it manages to invade the literary imagination?

There's a short poem by Earle Birney, called 'Can. Lit.', meaning 'Canadian Literature', which ends like this:

Eyes of Blood, Heart of Ice

We French & English, never lost
our civil war
endure it still
a bloodless civil bore
no wounded sirened off
no Whitman wanted
It's only by our lack of ghosts
we're haunted[1]

This poem dates from 1947, before there was a general acceptance among the intelligentsia of the notion that Canada had anything resembling a unique and viable inner life. Among other things, it's a very Canadian ritual genuflection towards the United States, from which all good things—in this poem, anyway—are supposed to come, including wounded men, poets, and ghosts. This poem continues to be quoted as a still-accurate encapsulation of the flatness of the Canadian imagination. The only problem with it is that it's not true; not of the literature, at any rate. Hordes of ghosts and related creatures populate Canadian fiction and poetry, from Robertson Davies through James Reaney to Timothy Findley; there's even an *Oxford Book of Canadian Ghost Stories*, edited by Alberto Manguel, who makes the provocative comment that most recent ghost stories in Canada have been written, not by the supposedly romantic francophones, but by the supposedly wooden-headed anglophones. 'In Québec and Acadia,' he says, 'literature in general and the ghost story in particular sprang from early legends and tall tales. . . . It is interesting to note that after those promising beginnings, French Canada has not been fruitful in ghosts. . . . "Our reality—political, physical—is perhaps too haunting in itself to allow for ghosts," the novelist Yves Beauchemin has said.'[2] And in real life, it was anglophone culture at its most yawn-inducing—that is,

[1] Earle Birney, *The Collected Poems of Earle Birney* (Toronto, 1975), i. 138.

[2] Alberto Manguel (ed.), *The Oxford Book of Canadian Ghost Stories* (Toronto, 1990), p. viii.

Toronto in the 1930s and 1940s—that produced Mackenzie King. King was the longest-reigning Prime Minister in our history and a walking synonym for dullness, but he was discovered post-mortem to have been governing the country through his dead mother, whom be believed had taken up residence in his dog. As one of Robertson Davies's characters remarks, 'Mackenzie King rules Canada because he himself is the embodiment of Canada—cold and cautious on the outside, dowdy and pussy in every overt action, but inside a mass of intuition and dark intimations.'[3] Boringness, in anglophone Canadian literature and even sometimes in its life, is often a disguise concealing dark doings in the cellar.

But Birney's 'lack of ghosts' line is even less true of that non-physical border, that imaginative frontier, where the imported European imagination meets and crosses with the Native indigenous one. If Birney had been gazing north instead of south—if he'd been looking even just a little way into his own backyard wilderness—he'd have found ample material to satisfy, not only his yen for dead people, but his longing for spirits as well. All that Robert Service rhetoric about the emptiness of the North, its vacancy, is meaningless when viewed in a Native light. For indigenous peoples the wilderness was not empty but full, and one of the things it was full of was monsters. The west coast forests are prowled, not only by the well-known Sasquatch, or Bigfoot, but by the Hamatsa and the Tsonoqua, to name just three. And in the eastern woodlands, there is the fearsome and many-faceted Wendigo.

Which brings us back to the initial question: what is the Wendigo? With a name like that, you'd expect it to have attracted the attention of the indefatigable American rhymester Ogden Nash, and this was in fact the case. By way

[3] Dunstan Ramsay, in Robertson Davies, *The Manticore* (Toronto, 1972), 99.

of introduction to the concept, here is Nash's 1936 poem 'The Wendigo':

> The Wendigo,
> The Wendigo!
> Its eyes are ice and indigo!
> Its blood is rank and yellowish!
> Its voice is hoarse and bellowish!
> Its tentacles are slithery,
> And scummy,
> Slimy,
> Leathery!
> Its lips are hungry blubbery,
> And smacky,
> Sucky,
> Rubbery!
>
> The Wendigo,
> The Wendigo!
> I saw it just a friend ago!
> Last night it lurked in Canada;
> Tonight, on your veranda!
> As you are lolling hammockwise
> It contemplates you stomachwise.
> You loll,
> It contemplates,
> It lollops.
> The rest is merely gulps and gollops.[4]

As you may just have gathered, the Wendigo is a cannibal. But that—and its location—is about all Ogden Nash got right. It's the Wendigo's heart, not its eyes, that's made of ice—the eyes themselves are supposed to roll in blood—and it has claws rather than tentacles. Although its voice bellows, it also whistles. Its lips are not blubbery, but blackened and eaten away, as if by decay, small animals, or frostbite.

[4] Ogden Nash, from *Bed Riddance: A Posy for the Indisposed* (Boston, 1969), 41.

Nash, being American, lived far enough away from the North so that he could reduce the Wendigo to an amusing poem; but for those who believe in it, the Wendigo is far from being a laughing matter.

In their indigenous versions, Wendigo legends and stories are confined to the eastern woodlands, and largely to Algonquian-speaking peoples such as the Woodland Cree and the Ojibway. The concept has many name-variations, including Weedigo, Wittako, Windagoo, and thirty-four others all beginning with W and having three syllables, as well as a number of forms beginning with different letters; all of these, and much else, have been catalogued by that compulsive collector John Robert Colombo, in his anthology of sightings, tales, and stories called *Windigo*.[5] But all agree that the Wendigo is—among other things—a giant spirit-creature with a heart and sometimes an entire body of ice, and prodigious strength; and that it can travel as fast as the wind. In some stories it has feet of fire, in others it makes tracks like giant snowshoes. It has no gender, although an individual Wendigo may once have been a man or a woman. It eats moss and frogs and mushrooms, but more particularly human beings; in fact, its prevailing characteristic seems to be its ravenous hunger for human flesh.

The literatures and folklores of many cultures contain man-eating giants, from the one-eyed Polyphemous of the *Odyssey* to the lumbering brutes of Grimms' fairy-tales to the overgrown English rhymester of 'Jack the Giant-Killer', with his murderous refrain. In the Western tradition, giants tend to be stupid foils for clever protagonists: David cunningly downs Goliath with his slingshot, Odysseus fools Polyphemous with a piece of word-play, Grimms' protagonists trick the giants into fighting with one another and escape with the magic boots or cloak, Jack is a wily sneak-thief who lives by his wits. Not

[5] John Robert Colombo (ed.), *Windigo: An Anthology of Fact and Fantastic Fiction* (Saskatoon, 1982), 2.

so in the stories of the dreaded Wendigo. You cannot outrun or outwit a Wendigo; most of the time you can't even converse with one, because the Wendigo has lost is capacity for human speech. Wendigos are very difficult to get rid of, though occasionally the task can be accomplished by magic. In some stories, influenced by cross-pollination with European werewolf tales, a silver bullet must be used. Once-human beings who have 'gone Wendigo' can be killed in the usual ways—knifing, strangulation, and so forth—but it is important to remove and burn the heart of ice, in order to melt it. Superficially, the heart of ice and the necessity of melting it might remind you of the glass sliver linked with narcissism in Anderson's *The Snow Queen*; but the love and tears proposed as a solution by Anderson don't do you any good at all *vis-à-vis* the Wendigo. Nor do the final arctic scenes in *Frankenstein* provide any clue to the Wendigo's meaning. Although the Frankenstein monster is gigantic, ugly, and destructive, he is much too articulate and theological to be a Wendigo.

Fear of the Wendigo is twofold: fear of being eaten by one, and fear of becoming one. Being eaten is simpler: a matter of mere gulps and gollops. Becoming one is the real horror, for, if you go Wendigo, you may end by losing your human mind and personality and destroying your own family members, or those you love most. You can be changed into a Wendigo by being bitten by one, or by tasting human flesh—even if driven to it by imminent starvation—or by being bewitched by a shaman. If you dream of a Wendigo, you are in great danger of turning into one yourself. Some people sing to the Wendigo, or make sacrifices to him, to ward him off. The Wendigo has been seen as the personification of winter, or hunger, or spiritual selfishness, and indeed the three are connected: winter is a time of scarcity, which gives rise to hunger, which gives rise to selfishness.[6]

[6] See ibid. 1–6.

The belief that one has become a Wendigo or has been possessed by the Wendigo spirit is a recognized form of insanity in the forested Canadian North, and has been thoroughly documented by Morton I. Teicher, in a paper called 'Windigo Psychosis among Algonkian-Speaking Indians'.[7] Says Teicher,

> The outstanding symptom of the aberration known as windigo psychosis is the intense, compulsive desire to eat human flesh. In many instances, this desire is satisfied through actual cannibal acts, usually directed against members of the individual's immediate family. . . . The individual who becomes a windigo is usually convinced that he has been possessed by the spirit of the windigo monster. He therefore believes that he has lost permanent control over his own actions and that the only possible solution is death. He will frequently plead for his own destruction and interpose no objection to his execution.[8]

Teicher calls this, with admirable understatement, 'a clear and severe case of psycho-social dysfunctioning'. He seeks to show how, in a subsistence culture in which food is hard to come by and starvation threatens frequently—and in which group effort, the suppression of individual emotions, food-sharing, and, incidentally, the abhorrence of cannibalism, are stressed—the Wendigo belief allows one to violate numerous profoundly held taboos, and provides an explanation for these violations that places the acts outside one's own control. 'The Devil made me do it' of fundamentalist Christianity is replaced by 'The Wendigo made me do it.' The clinical symptoms of this disease include anorexia or nausea, the abhorrence of all food except the longed-for human variety, and 'deep depression . . . to the point of stupor', followed by episodes of violence.[9] Teicher is interested in the extent to which belief determines behaviour, and concludes that, in the

[7] John Robert Colombo (ed.), *Windigo: An Anthology of Fact and Fantastic Fiction* (Saskatoon, 1982), 164.

[8] Ibid. 167, 168.

[9] Ibid. 169.

case of windigo psychosis at any rate, belief is the controlling factor: if you believe in Wendigos, you are capable of believing that you yourself have turned into one, and of acting accordingly.

This then is the Wendigo, in both its evil-spirit form and its human variations. Since the arrival of paper in North America, this creature has made frequent appearances on the page, beginning with notations in explorers' journals. More recently, it has attracted the attention of a number of literary writers, not all of them Canadian: Algernon Blackwood, for instance, added it to his gallery of scary monsters in his short story 'The Wendigo'. But when the Wendigo gets into this kind of written material—especially when the material is produced by non-Natives—what is it that the Wendigo is understood to be doing there? What are its metaphorical functions, in relation to the rest of the elements in the story? What purposes is it made to serve?

In short and essentially lyric poems, the Wendigo can be an object of meditation, much as, say, a seventeenth-century skull, a nineteenth-century daffodil, or a twentieth-century slug or skunk. You would think it would fit in well, also, with the late twentieth century's project of humanizing traditional monsters, as exemplified by, for instance, John Gardner's rewriting of Beowulf from the point of view of Grendel, or Ann Rice's exaltation of the vampire to existential hero. But, since one of its characteristics is its absence of language, it is difficult to show a Wendigo as a speaking or even a thinking subject. The Wendigo has tended to remain an object, and to be slanted in the opposite direction: it's not that monsters are human, but humans themselves are potential monsters. The Wendigo is what you might turn into if you don't watch out.

Watch out for what? In George Bowering's poem 'Windigo', the vulnerable spot is the heart. His poem begins with description, and he covers the traditional features:

Windigo
is twenty-five feet tall,
a long shadow
on the ice

& ice in him,
his heart,
made of hard ice.

He lives in the forest
of thin dark trees, north
where the sun
reaches on a slant,
casting long shadows,
    the sun is too far
    to reach that heart
    of ice,
    to melt it.[10]

The poem goes on to describe the Wendigo's visual and dental problems, eating habits, and bad behaviour generally, in considerable detail: Bowering has obviously read a lot of Wendigo material. At the end, however, he turns to the Wendigo's relation to the rest of us: what can be done about the Wendigo should one turn up in your vicinity, how we should comport ourselves towards it, and why:

    To kill
Windigo, a silver bullet
into the heart of ice, or
the shaman's glassy stare
over his flames, no human
heart of blood can bring
        that mountain to the ground.

[10] George Bowering, *The Gangs of Kosmos* (Toronto, 1969), 28.

Eyes of Blood, Heart of Ice

Windigo,
                they say he was
once a man, winter
entered him, the wind
from the north in artery
to his heart, that ice, flow
thru his changing body.

. . .

So sing to the Windigo
for mercy. Sing the shaman's song,
place food & drink, sacrifice
to the brother made beast,
the Windigo could be your brother,
he walks the forest at night,
screaming for your flesh,
sing the sacrifice song.

The song of the hunter,
the scream of the Windigo,
the heart of ice
                appears
in the chest of the man
who suddenly craves a man's flesh.
He must be a sacrifice too,
he must ask to be sacrifice,
he could be the Windigo of his brother,
his song can turn to scream of the ice heart.

. . .

Come home, hunter,
before nightfall,
your brother is Windigo.
The heart of ice
was heart of blood,
winter
has entered him.

> So sing to the Windigo for mercy,
> your brother,
> sing the shaman's song,
> & sing for your own heart too.[11]

Bowering's emphasis is on the relationship between Wendigo and human—the creature could be your brother, and your 'own heart' is in danger of becoming, like his, ice-cold. Praying to the Wendigo for mercy in oxymoronic—the Wendigo by definition is incapable of mercy—but what Bowering seems to be getting at is that we ourselves should take mercy on the Wendigo, and thus on our own all-too-human—and therefore potentially hard and icy and monstrous—hearts.

Paulette Jiles takes a tougher line in her poem, which is also called 'Windigo'. She assumes more basic Wendigo knowledge in the reader than Bowering does: she doesn't spend a lot of time on the Wendigo's heart and its food preferences, but turns quickly to an exploration of its possibilities as an emblem standing for aspects of the spiritual life. Here is the poem:

> No one understands the Windigo, his voice like
> the white light of hydrogen, only long.
> Some say he carries his head under his arm, for
> others it is the race down to the rapids
> where the canoes draw close, close to the shore
> and he jumps in. You have time for a few last words.
>
> Under the moon he turns pale grey, the
> head chatters amiably about meals. He is the
> Hungry Man, the one who reached this wasteland of
> the soul and did not emerge. Not whole. Not as you
> would recognize wholeness.

[11] George Bowering, *The Gangs of Kosmos* (Toronto, 1969), 32, 33.

Sometimes he wants to be killed, putting his
heart or what there is of it in the way of arrows,
bullets, he wants his soul or what there is of it
to spring heavenward to the village where people
begin again, he too
wants to cross the bridge.
His story is of one who reached starvation
and death and did not make it through, not
as you would recognize making it.
People shoot the Windigo, they
do not pray for him, or it.[12]

No pleas for mercy here. The Wendigo is placed firmly on the other side of the pale; by the end of the poem his is not even a 'him' but an 'it'. Still, he is a fellow traveller: what we have in common with him is the journey through the 'waste-land of the soul', but where we part company is that we are assumed to be survivors, so far, and he is not. If Bowering's poem has a Catholic feeling about it—prayers for lost souls—Jiles's is firmly *Pilgrim's Progress.* Like all those fallen by the wayside on the road to the Heavenly City, the Wendigo has definitely not made it through. He brings out in us, not the impulse towards mercy, but our own will to survive with some kind of integrity or 'wholeness'. Despite his muddled longings for some kind of salvation, the Wendigo is a psychic fragment; and, as we will see, psychic fragments are dangerous to the health.

When Wendigos turn up in longer literary narratives, they tend to function much in the way of ghost functions in the ghost story or Gothic tale. In such stories—to generalize wildly—the other-worldly creatures may exist in one of three relationships to the human characters in the story. First, such a being may be presented as a manifestation of the environment; the spirit

[12] Paulette Jiles, *Celestial Navigation: Poems* (Toronto, 1984), 104.

is a spirit of place. If you go into a haunted house, you will be haunted. The haunting has nothing to do with you, or with your past or inner life: it's your presence in the infested location that triggers the experience. A great many ghost stories, especially the kind kids tell around campfires, are like this, and so are many Wendigo stories. If you go into the woods tonight you're in for a big surprise, but your sad fate will not be your fault. In these stories, meeting up with a Wendigo carries no more moral weight than meeting up with a bear. If you get eaten, about all that can be said of you is that you ought to have been less unlucky.

In the second kind of story, the ghost, spirit, or creature has a message for the protagonist—as Hamlet's father's ghost does—or comes to reward you, to consummate a love relationship—as in those mournful folk-songs about dead lovers returning all dressed in white—or to punish you for something you've done. The creature has a connection with you: with your past behaviour, your spiritual status, your family position. It's the bearer of a fate which the logic of the story requires you to live out. If you get eaten by a Wendigo of this second sort, we don't feel sorry for you, because it was definitely your doing. In fact we feel protected, because as long as we behave ourselves the Wendigo will leave us alone.

In the third kind of story, the other-worldly creature is a fragment of the protagonist's psyche, a sliver of his repressed inner life made visible. The purest example of this kind of ghost is in Henry James's 'The Jolly Corner', and one of the most subtle is in 'The Turn of the Screw'. Wendigoes of this third sort are likely to be human beings who have 'become Wendigo', who have turned themselves inside out, so that the creature they may only have feared or dreamed about splits off from the rest of the personality, destroys it, and becomes manifest through the victim's body. These tales are tales of madness, and link up, circularly, with the first kind—the spirit of place takes possession of you, causing a part of you to externalize itself. No one 'goes Wendigo', it seems, in the middle of a city. The bush, the trees, the loneliness, are essential.

Wendigoes, with their singularly sub- or non-human nature and their terrifying mindlessness, lend themselves better to stories of the first and third types than to the second. Spirits of the second type have communications to make, and Wendigoes, devoid of language, are very bad at communicating. But the Wendigo has in fact been used by writers for all three purposes.

I encountered my first Wendigo story when I was an adolescent, and it scared me silly, which is what stories of the first type are supposed to do. It's in a book called *Brown Waters and Other Sketches*, by W. H. Blake, and it dates from 1915. *Brown Waters* does not purport to be fiction. It's a collection of that genre beloved of the nineteenth century in general and Washington Irving in particular, the 'sketch', composed of supposedly real-life landscape description worked up in an artistic fashion and interspersed with accounts of local tales. This particular 'sketch' from *Brown Waters* is called 'A Tale of the Grand Jardin'. But instead of being set in picturesque Europe or the more civilized parts of east coast New England, it's located in the Canadian North, 'those great barrens that lie far-stretching and desolate among the Laurentian Mountains', and the influence is less that of the whimsical Irving than of the Gothic Edgar Allan Poe.

The tale is a story within a story. During a thunderstorm, the narrator's fishing companion tells him about something that happened to him on a similar excursion. First Blake sets the scene:

So were we two alone in one of the loneliest places this wide earth knows. Mile upon mile of gray moss; weathered granite clad in ash-coloured lichen; old *brûlé*,—the trees here fallen in wind-rows, there standing bleached and lifeless, making the hilltops look barer, like the sparse white hairs of age. Only in the gullies a little greenness,—

dwarfed larches, gnarled birches, tiny firs a hundred years old,—and always moss . . . great boulders covered with it, the very quagmires mossed over so that a careless step plunges one into the sucking black ooze below.[13]

Shades of the tarn beside the House of Usher. If an elemental spirit is going to be exuded by this landscape, you know it won't be a pixie. And, what with all that moss, anyone who knows the eating habits of the Wendigo when there are no human beings available should feel an immediate pricking of the thumbs.

The narrator's companion proceeds to describe the earlier disappearance of a guide named Paul Duchêne—a man very familiar with the wilderness, and of an almost 'demonic' energy, unlikely to become lost—who has gone 'out of his mind' and has wandered off, never to reappear. He then mentions a belief of the local Montagnais Indians, a

strange medley of Paganism and Christianity, that those who die insane without the blessing of a priest become wendigos,—werewolves, with nothing human but their form, soulless beings of diabolical strength and cunning, that wander for all time seeking only to harm whomever comes their way. A black superstitious race these Indians are, and horribly sincere in their faith. They shot down a young girl with the beads of her rosary, because her mind was weakening, and they thought thus to avert the fate from her, and themselves.[14]

The previous summer, he continues, he was travelling in an area called the Rivière à L'Enfer, where he camped beside a lake with black water. His guides went off for supplies, leaving him alone, whereupon he experienced a strange oppression of the spirit. 'In what subtle way', he asks, 'does the universe convey the knowledge that it has ceased to be friendly?' That night a tremendous storm blew up. Sitting in

[13] W. H. Blake, *Brown Waters and Other Sketches* (Toronto, 1940), 92, 93.
[14] Ibid. 95.

his tent, he heard an unearthly cry, which was 'not voice of beast or bird'. He burst from the tent and was confronted by a creature—'something in the form of a man'—which sprang to him. He managed to escape only by dividing into the late. 'And what in God's name was it?' asks our narrator. The storyteller replies, 'Pray Him it was not poor Duchêne in the flesh.'[15]

This Wendigo is a human being who has been claimed by the sinister wilderness and has become an expression of it. Incidentally, by one of the laws of the Wendigo, whereby those who see the Wendigo *become* the Wendigo, the tale-teller himself ought to have been in jeopardy. In order to be consistent with the mythology, the sketch should have ended with the fishing companion emitting a fearful whistle and taking a bite out of Blake's neck. But Blake the writer was far enough removed from the beliefs he incorporates into his story to leave the Wendigo as a sort of stage-effect: the terrifying conclusion to an Edwardian tale built largely on the atmosphere of its locale. 'Poor Duchêne' does not deserve his fate. He has not been a bad man, or even an incompetent one. He is simply a victim of landscape.

Not so Cyprien Palache, the ill-starred foreman of a logging gang in William Henry Drummond's narrative poem 'The Windigo' (1901). Drummond wrote a great many poems, some of which were still around in readers when I was attending school, though mostly they were passing out of favour—probably because they were written in a sort of Franglais patois which was supposed to represent 'French Canadian' speech. This kind of dialect now makes good liberals cringe, but not so in the nineteenth century; lots of poets wrote in dialect, including Tennyson and especially Kipling, and even so redoubtable a Québécois as Louis Frechette—who wrote an introduction to Drummond's 1926 *Complete Poems*—

[15] Ibid. 100.

applauded Drummond's efforts to present his jolly, likeable, superstitious, down-to-earth *habitant* characters to an anglophone audience.

'The Windigo' is a supernatural story of the second type, so the bad thing that happens to Cyprien Palache is a direct result of the bad thing he himself has been doing to someone else. Palache has tiny eyes—always a bad sign—he swears, and there's a suggestion that he's already made a pact with the Devil, getting, in return, the traditional musical instrument, although it is not the favoured European violin. Here's his first appearance:

> Beeg feller, alway watchin' on hees leetle weasel eye,
> De gang dey can't do not'ing but he see dem purty quick,
> Wit' hees 'Hi dere, w'at you doin'?' ev'ry tam he's passin' by
> An' de bad word he was usin', wall! it often mak' me sick.
>
> An' he carry silver w'issle wit' de chain aroun' hees neck
> For fear he mebbe los' it, and ev'rybody say
> He mus' buy it from de devil w'en he's passin' on Kebeck
> But if it's true dat story, I dunno how moche he pay.[16]

But this isn't the worst. Palache is also a sadistic racist. The lumber gang has come by an orphaned Indian boy, left nearly starving when his parents have drowned doing a very un-Indian thing, namely, trying to shoot a dangerous rapid during spring spate (Indians traditionally portaged if there was any doubt: daredevil white-water canoeing is an invention of whites). But orphanhood for Johnnie is essential to the plot, so Drummond bumps off the parents. Palache treats Johnnie brutally, and blows his little silver whistle whenever he wants to get his hands on him. Luckily, the other men and the cook take his part in secret, and keep him from starvation.

However, Johnnie is not without recourse. Says the narrator,

[16] William Henry Drummond, *Dr W. H. Drummond's Complete Poems* (Toronto, 1926), 178, 179.

I see heem hidin' somet'ing wan morning by de shore
So firse tam I was passin' I scape away de snow
An' it's rabbit skin he's ketchin' on de swamp de day before,
Leetle Injun Johnnie's workin' on de spirit Windigo.[17]

His sacrificial offerings get the desired result. It's winter, season of old-time logging and also of Wendigoes. The narrator and Cyprien Palache are returning from an exploratory hunt for new timber, on their snow-shoes, when Cyprien loses his whistle. He kicks up a huge fuss about it, and sends Johnnie the orphan out in the stormy night, telling him he can't come back until he finds it. Such an order, on such a night, means almost certain death. The men wait in the shanty, unable to sleep, as the storm builds outside and the atmospheric effects heighten. They start hearing strange noises, and one of them sees a face at the window: 'An' it's lookin' so he tole us, jus' de sam' as Windigo.' At last there comes a fatal knock at the door, and the call of the whistle: the silver whistle, certainly, but also—perhaps—the characteristic whistling sound made by Wendigoes. Cyprien, summoned by his own diabolical whistle, rushes out in the storm, never to be seen again. In the morning, after the storm, the men discover a giant snow-shoe track outside the cabin, 'bout de size of double sleigh | Dere's no mistak' it's makin' by de spirit Windigo.' Johnnie, on the other hand, returns to his hunting and trapping ways, and survives and prospers. Evil has been punished, justice has been done.

It's the end of the story, but not of a typical Wendigo story. Mostly, Wendigoes are not so biddable, even in return for rabbit skins, even by wronged orphans, and as a rule they show no interest at all in whether you've been bad or good. This Wendigo resembles the Christian devil who carries off Dr Faustus more than it does the majority of its Wendigo relatives.

[17] Ibid. 181.

The third type of story is the one in which the supernatural creature represents a split-off part of the protagonist's psyche, and in which insanity therefore plays a central role. The North, as we've seen, has frequently been credited with driving strong men mad, and among Wendigo believers, who have tended to be Native people, this madness may take the form of thinking you are turning into a Wendigo. For whites who go north and go crazy too, there are other forms, and names for them as well. 'Cabin fever' is one; 'bushed' is another. Earle Birney has a well-known lyric poem called 'Bushed', in which a man who has attempted a life of solitude within nature starts feeling that a nearby mountain has come alive in unpleasant ways and is out to get him.[18] Robert W. Service, that great mother-lode of Northern cliché, provides a number of Northern crazies, such as the raving narrator of 'The Ballad of Pious Pete',[19] whose saga is introduced with the terse epigram '"The North has got him."—Yukonism.' Then there's a very heroic poem about Clancy of the Mounted Police, an 'unconscious hero of the waste', who obeys the injunctions of his 'little Crimson Manual' where it's 'written plain and clear | That who would wear the scarlet coat shall say good-bye to fear'. Clancy responds to a call to rescue a 'White man starving and crazy on the banks of the Nordenscold', and Service sums up the northern-madness situation in a few well-chosen words:

> Cold with its creeping terror, cold with its sudden clinch;
> Cold so utter you wonder if 'twill ever again be warm;
> Clancy grinned as he shuddered: 'Surely it isn't a cinch
> Being wet-nurse to a loony in the teeth of an arctic storm.'[20]

The most famous real-life case of this kind of northern craziness is a man known as the Mad Trapper of Rat River.

[18] Birney, *Collected Poems*, i. 160.
[19] Robert W. Service, *The Complete Poems of Robert Service* (New York, 1945), 98.
[20] Ibid. 156.

In the 1930s he shot—for no apparent reason—a Mountie who had approached his isolated cabin; then he led a whole team of Mounties on an intricate chase, on foot, across impossible terrain, before he was finally hunted down with the aid of a First World War flying ace. No one ever found out why the Mad Trapper was mad, or even who he really was. Novelist Rudy Wiebe has written about the Mad Trapper twice, once in a short story, 'The Naming of Albert Johnson',[21] and again in a full-length novel, *The Mad Trapper*, but he shies away from ascribing motive and from the inner gibbering that would have tempted Service, presenting the man instead, through his incredible feats of endurance and his entire lack of interest in human language, as a monstrous but sympathetic cross between a superhuman and a subhuman.

White people and Native people have been interacting in the North, and cross-pollinating one anothers' inner landscapes, for hundreds of years now, and the concepts of getting 'bushed' and 'going Wendigo' can overlap in interesting ways. The first supposes that the craziness comes from inside the self, the second that it originates from outside, in the form of possession by a spirit; when the two blend, the reader is free to choose. In such hybrids the result is likely to resemble a story of the third type, and the Wendigo—or Wendigoization, to coin a term—tends to suggest a split-off element of the protagonist's psyche, which develops a life of its own and takes him over.

This is a subtext of Wayland Drew's complex 1973 novel *The Wabeno Feast*. The book has a triple time-line: first, a future in which ecological catastrophe has arrived in full spate, due to modern white-man's greed and lust for power; second, an earlier canoe trip in the past lives of the same characters, in which they try to play explorers and get woefully lost, like Franklin, due to their arrogance and their inadequate

[21] Rudy Wiebe, *Where is the Voice Coming From?* (Toronto, 1974), 145.

knowledge of both technique and environment; and third, the journal of an early nineteenth-century man named MacKay. MacKay is a priggish rationalizer who has journeyed from Scotland to take up a post as clerk to the Hudson's Bay Company in the wilds north of Lake Superior, and is possibly mad before he even sets foot on Canada, though his later wilderness experiences don't help any. His literary ancestry may go back to Mr Kurtz, of *Heart of Darkness* renown, or, perhaps, to those other deranged Scots, the protagonists of Hogg's *Confessions of a Justified Sinner* and Stevenson's *Dr Jekyll and Mr Hyde.* In any case, MacKay too has a sinister *alter ego*, a man called Elborn, who follows him around, mocking at his plans and pretensions, which include a slavish devotion to the exploitative mercantile aims of the Company and entrenched feelings of superiority and contempt for the Native people with whom he comes in contact. Later, we realize that Elborn probably doesn't exist, but is a symptom of MacKay's split inner self. In killing him—as he plans to do at the end—MacKay, like Poe's William Wilson, will kill himself.

It's in MacKay's journal that the Wendigo makes its appearance. In fact, MacKay encounters two Native creatures beginning with W, a letter which you have surely come to regard with some suspicion. Early in his journey into the hinterlands, he hears about the Wendigo from a group of voyageurs, who are encouraged by Elborn, or the Elborn side of MacKay's personality, to spin 'their tales of death and terror'.

They tell of the wendigo, a mythical creature of Indian lore, and each elaborates on the other's imagining until in his fantasy Elborn beholds a creature thirty feet in height, a naked, hissing demon whose frog-like eyes search out unwary travellers and roll in blood with craving to consume them! Another whispers that the creature lacks lips to cover its shattered teeth, and a third describes its feet like scabrous canoes on which it rocks howling through the swamps at evening.[22]

[22] Wayland Drew, *The Wabeno Feast* (Toronto, 1973), 23.

MacKay comments, 'They are children who listen to such stuff!' This is of course the wrong attitude, one of the first laws of late twentieth-century Canadian literature being that he who scorns Native beliefs will come to a sticky end.

The second W-word that enters the story is *Wabeno*, which is explained at greater length. The Wabeno is a sorcerer of immense power, and MacKay witnesses a feast conducted on an island by a Wabeno and his followers. These Wabenos are taller than usual and whiter than usual, owing to the fact that they have burnt their skins repeatedly by dancing in fire. This same activity has seared off their vital parts, and the watching MacKay comments, 'I supposed I beheld a dream, a nightmare wherein fiends of neither sex made their mockery of both.'[23] The Wabenos are also cannibals, and the food at the 'Wabeno Feast' of the title consists of one of their own members. The feast itself concludes with the whole island in flames.

The other Indians hate and fear the Wabenos, and hope, now that the Company has come among them, that the influence of the Wabenos will diminish. But in fact the Company and the Wabenos stand for the same thing, the destructive lust for power over others. The Company, after all, makes its money by trading alcohol and guns for beaver skins, and all the Indians who trade with the Company and live near it are demoralized and debauched. MacKay begins to realize this when he encounters a tribe which has refused to deal with the Company at all. These people, unlike the others, are happy and healthy, and MacKay falls in love with one of the women and leaves the Company to be with her, pretending to himself that he's forming the alliance in aid of future trade. This Native woman tries to do the traditional literary thing, that is, she tries to save the white man from the consequences of his own rashness and folly. But MacKay resists her attempts to teach him how to live in harmony with nature, and sets about

[23] Ibid. 87.

chopping down trees to build a log cabin. When the woman protests that such a house would be like a cage, he orders her to gather food. She brings him a little edible message, which conveys her opinion of what he is becoming: a basket of toadstools. Wendigo food. At this point MacKay loses control entirely, snarls 'like a cornered wolverine', shouts 'gibberish' at her, and does his best to rape her. Needless to say, she doesn't stick around for any more of this, and MacKay is left alone, to live in his log fortress and to develop further symptoms of cabin fever. He descends into apathy, depression, and near-starvation, until Elborn again materializes to taunt him and to set off his final symptom, which is murderous paranoia.

The Wendigo, in Wayland Drew's novel, is what you will become if you try to be a Wabeno: that is, the desire to be superhuman results in the loss of whatever small amount of humanity you may still retain. The Wendigo is also a sort of gatepost, one of a pair of emblems which mark the poles between which the various protagonists move. But above all, the Wendigo is a part of MacKay which he himself has suppressed and denied. It is his own emotional life, turned rabid, seeking only to harm whomever comes in his way. One of Elborn's last jibes at MacKay is, that without him, MacKay could have forgotten that he was 'an animal, with the passions of an animal'. Poor MacKay replies, 'I shouted that I was not an animal. That I was a man.'[24]

The man–animal dichotomy is illustrated by the character Angus—another demented Scot—in Graeme Gibson's novel *Perpetual Motion.* Angus is a Wendigo in most respects, although he is given, for a change, a positive role. He runs away into the forest in horror and outrage over the slaughter of Passenger Pigeons conducted by his father for money, and he turns wild. But then, there was something wild about him

[24] Wayland Drew, *The Wabeno Feast* (Toronto, 1973), 264.

from the beginning: he was born with a full set of teeth, and covered with hair, and as a child he has spent a week in a bear's den, being mothered by the bear, like the children in many a European and North American wild-child tale. After he goes wild, Angus loses language, and emits only howls, moans, and whistles; he grows hair, acquires the incredible speed and strength of the Wendigo, and loses his previous personality: 'Stripped of all particulars, his face burned like a guttering candle until his eyes turned dead as snow.'[25] The other characters in the book festoon him with Wendigo ice-images during their speculations about him. The hunters think he may have gone 'bushed': 'Touched by the wild', they say, in Robert W. Service parlance. 'It got into him like ice.' 'Froze his soul.'[26] However, because the nature–civilization polarities in the novel have been violently reversed, with nature and animals seen as good and human beings, especially technologically minded ones, seen as hideously bad, Angus is a Wendigo minus the evil. He is not a cannibal. Nor is he a fragment of his own repressed psyche, but a fragment of his hated father's. His mother gets it right: 'Angus was a mirror.'[27] What he reflects is the hurt being inflicted, by human beings, to the wilderness, and the hurt they are therefore inflicting on the wild part of themselves. The real dangerous lunatic in *Perpetual Motion* is not Angus, but his rationalistic father, Robert Fraser.

All of the Wendigos examined so far have been men, and all except one of the Wendigo texts I've spoken about so far have been written by men. But, as we know, there can be female Wendigoes too. What happens when a woman writer gets her hands on the Wendigo and refurbishes it as a metaphor, not for selfishness or greed or hunger of a material nature, but for sexual passion? This fascinating subject, among others, will be examined in my next lecture, which is about

[25] Graeme Gibson, *Perpetual Motion* (Toronto, 1982), 231.
[26] Ibid. 248.
[27] Ibid. 250.

women rewriting wilderness. Meanwhile, I hope you have all learned enough about Wendigoes so you will be able to avoid being devoured by your relatives, in any but a figurative sense.

4
Linoleum Caves

In my first three lectures, I dealt with three clusters of story and image relating to the Canadian North. The first one was about the Franklin expedition, and the versions of the North surrounding it; the second was about men who wanted—like Grey Owl—to turn themselves into Indians. And the third was about how the Wendigo—a giant cannibalistic ice-hearted Algonquin Indian monster—has been used by non-Native writers.

But one thing must have occurred to you as I've gone along: the North, as we've seen it described so far, has been inhabited primarily by men. 'There are strange things done in the midnight sun | By the men who moil for gold,' say the opening two lines of the third most famous poem in Canada, Robert Service's 'The Cremation of Sam McGee'.[1] But what happens if it's *women* who are moiling for gold, or do women moil, and if they do moil, is it likely to be for gold, and if so, is their moiling of the same nature as male moiling? Possibly they moil for males; if so, is the female moiler of the species more deadly than the male moiler? Ought we to be speaking of femoilers, or of gun moils? Or is all of this merely

[1] Robert W. Service, *The Complete Poems of Robert Service* (New York, 1945), 33; the first and second most famous are 'In Flanders Fields', by John McCrae, and Service's 'The Shooting of Dan McGrew'.

rhetorical, since there are so few women in the North of any sort, moiling or otherwise?

The title of this lecture is 'Linoleum Caves', which comes from a sentence in Alice Munro's 1971 novel *The Lives of Girls & Women*: 'People's lives, in Jubilee as elsewhere, were dull, simple, amazing and unfathomable—deep caves paved with kitchen linoleum.'[2] What struck me about that last phrase was the juxtaposition of 'caves' and 'linoleum'—the idea of domesticity as simply a thin overlay covering a natural, and wild, abyss. Or, conversely, the suggestion that you can pave wilderness over, make it into a kitchen, however thin the linoleum veneer. Now I would like to examine what happens to the North—that is, the written North—the North both of clichéd image, and of more serious literature—when *women* enter the northern landscape, either as authors, or as female or male protagonists created by women authors. What new possibilities are there, for outrage, treachery, salvation, and refuge, or merely harmless play, when women get their paws on those mainly manly Northern icons?

And—looking at the subject from another angle—what happens when you take works by Canadian women writers—which are usually read in the context of women's writing in general—and read them instead, or in addition, in another context—that of the body of Northern imagery and lore of which they may also form a part? It is such questions that will trouble our waking hours this afternoon.

In my first lecture, I talked about the Franklin disaster, and the complex of imagery and story that has accumulated around the idea of the North as a mean female—a sort of icy and savage *femme fatale* who will drive you crazy and claim you for her own. But what happens if the 'you' that is being driven and/or claimed is not a man, but a woman? What happens then to the idea of the North as female, and what sorts

[2] Alice Munro, *Lives of Girls & Women* (Toronto, 1971), 253.

of relations, sexual or otherwise, are created by the juxtaposition of subject and landscape? Does the land become male? Or does it become neuter, or else a wicked stepmother, or a rival in love? Or does it stay female, but turn benign, a kind mother or sister or lesbian lover? What bundles of images and associations, what suitcases full of ready-made attitudes, do you have to carry, or else discard, if you make this attempt?

Before we start rummaging through the undergrowth for tell-tale gender signs, let me tackle a question I've been asked a number of times: why are there so many Canadian women writers?

Sometimes this question is asked accusingly, or plaintively, or despairingly, as in 'Why are there so many mosquitoes?' For these questioners, *some* is too many. There are only 'some', by the way. There are not, for instance, half. But everything is relative, and relative to, say, Egypt or Wales, there are indeed a lot of well-known Canadian women writers.

I don't know the answer. But I can hazard some guesses. One possible reason may be that, for those seeking role-models and literary ancestresses, Canada provides quite a few. A fair share of the first writing done in Canada was done by women—in Québec by the French nuns who came to do missionary work, in anglophone Canada by the wives of the middle-class or younger-son settlers forced out of Britain by unemployment and hard times after the Napoleonic wars. The poet who is usually cited as striking the first 'authentic' lyric note in nineteenth-century poetry is a woman—Isabella Valency Crawford. Pauline Johnson was the best-known Canadian poet at the turn of the century, and in the generation preceding mine, many of the most respected poets were women, as was the pre-eminent novelist, Margaret Laurence.

What this boils down to is that whatever you may think of them, Canadian women writers cannot be *avoided*—as women writers were, for years, in the States, where Melville, Hawthorne, Hemingway, James, Fitzgerald, Faulkner, and the boys were placed firmly centre-stage, at the expense of writers such as Edith Wharton and Willa Cather. Maybe part of the answer is that there were no towering, overpowering *male* Canadian literary figures, so there was room for women. Or maybe, under frontier conditions, the men were kept so busy chopping down trees and strangling wolves that the arts came to be regarded as sissy stuff, and women were left to do the cushion embroidery, the flower-painting, and the poetry-writing. Or perhaps being so close to frontier conditions, under which women with lots of muscles were valued for their water-hauling and plough-pulling capabilities, Canadians never developed the concept of women as merely brainless decoration. Canadian oral folklore is still full of tales of our grandmothers' generation, when women ran farms, chased off bears, delivered their own babies in remote locations and bit off the umbilical cords. Whatever the reasons, the fact remains: if you're looking at writing in Canada at all, you can't just footnote the women.

This is true despite another truth—that large areas of Northern mythology are practically devoid of women. For instance, there are no stories about female explorers, which is perhaps linked to the absence of female explorers in real life. Women, when they appear in male explorer stories, are not explorers themselves, but explorees: wives of the Natives, features of the newly discovered terrain. The Robert W. Service North of popular image is assumed to be a man's world; even though the North itself, or herself, is a cold and savage female, the drama enacted in it—or her—is a man's drama, and those who play it out are men. There are no Robert Service women mushing their dog teams, staking their claims, being driven crazy, and freezing to death. There *are* some women in Service's Yukon, of course, but they are not protagonists; they are ladies of the night, like the thievin', cheatin', seductive 'Lady Who's Known as Lou' of Dan McGrew fame,[3] or down-

[3] 'The Shooting of Dan McGrew', *Complete Poems*, 29.

trodden and debaunched Native women, and the occasional pure, sweet wife who exists to be abandoned when the lure of the North gets too much and the husband goes off to do the required mushing, prospecting, and freezing to death. The women in Service are not incarnations of the true North, wild and free: the land itself gets to hog that role alone. So I can't offer you any female Franklins—no one would have funded an expedition headed by such a person—or any female prospectors, raving mad or otherwise, or even any female Mounties, out to get their man—though some in that category may come along soon.

There's also a shortage of stories about women who wanted to turn themselves into Indians, like Grey Owl, and the reason for this has been, traditionally, not far to seek: boys who ended up that way were inspired largely by Fenimore Cooper and his tales of frontier warfare, and these offered no really fulfilling roles for girls, since Indian women were not warriors. Even in Ernest Thompson Seton's more peaceable kingdom, women were, to put it bluntly, deeply secondary. Pretending you're part of a hunter-gatherer society is not much fun if all you get to do is the gathering. Now that gathering has been somewhat upgraded, however, there may be scope in the future.

Not that women haven't written *about* Indians. Many have, or have incorporated Indian motifs into their work. Try Susan Musgrave and Sheila Watson, for starters. But the notion of actually *becoming* an Indian does not seem to have had much appeal. About the nearest thing I can come up with for you is E. Pauline Johnson, who flourished in the last decades of the nineteenth century. She was a poet—whose poem 'The Song my Paddle Sings',[4] was standard fare in Canadian school readers until the 1950s. She was also what would now be known as a performance artist, but was called at the time an

[4] Pauline E. Johnson, *Flint and Feather: The Complete Poems of E. Pauline Johnson (Tekahionwake)* (Toronto, 1969), 31.

elocutionist—that is, she gave public performances in drawing-rooms and theatres, at which she recited her own poetry to great dramatic effect. Her performances were divided into two parts—the first, in which she wore an evening gown and recited her more conventional poems; and the second, the 'Indian Princess' half, for which she put on a buckskin-and-fur costume she'd designed herself, and curdled the blood of her audiences with a more fangy and toothy variety of verse. Pauline Johnson was in fact only about ⅜ Mohawk, and she had been brought up in strict Victorian fashion by her white mother. However, she always identified herself as Mohawk, and idealized her mostly Mohawk father. The 'Indian' half of her poetic persona allowed her a great deal more freedom than did the 'white' half: she could use it to give expression to sentiments that were frowned upon for Victorian women, as well as to do a certain amount of advocacy pleading for the Indians. In 'Ojistoh', for instance—one of her most effective performance pieces—a Mohawk woman kidnapped by a Huron warrior steals his knife and stabs him to death with a good deal of relish;[5] and 'The Cattle Thief' is astonishingly modern in its reproaches to the whites:

What have you left to us of land, what have you left of game,
What have you brought but evil, and curses since you came? . . .
Go back with your new religion, we never have understood
Your robbing an Indian's *body*, and mocking his *soul* with food. . . .
Give back our land and our country, give back our herds of game;
Give back the furs and the forests that were ours before you came . . .[6]

Pauline Johnson was also outspoken about the handling of Indian women by writers. She wrote a long critical piece for the *Toronto Sunday Globe* of 22 May 1892 on 'The Indian Girl in Modern Fiction', in which she tears strips off a number of

[5] Pauline E. Johnson, *Flint and Feather: The Complete Poems of E. Pauline Johnson (Tekahionwake)* (Toronto, 1969), 3.

[6] Ibid. 12.

white authors for dishing up, again and again, the same kind of Indian maiden in their books—a poor, doomed creature, who passionately loves the white hero—like Oucanasta in Richardson's *Wacousta*—but is not loved in return, who is 'so much wrapped up in him that she is treacherous to her own people, tells falsehoods to her father and the other chiefs of the tribe, and generally makes herself detestable and dishonourable', who is 'dog-like, fawn-like, deer-footed, fire-eyed, crouching, submissive'—all these are adjectives she quotes from real books—and who usually ends her life by suicide, because of the perfidy of her white lover.[7] Perhaps in retaliation, Johnson herself wrote a haunting poem in which it's the white man who dies for the Indian woman instead—not suicide, but a death by blizzard—though, to be fair, she polishes off the girl also when she goes out to find him. The poem is 'The Pilot of the Plains'; it too was often in school readers, and I'm willing to bet it influenced more than one writer of Canadian ghost stories—more than one, that is, besides me. I don't know whether I can do it justice, but what you should imagine is Pauline Johnson herself, or Tekahionwake, which was her Mohawk name, with her hair down and a bear-claw necklace around her throat, giving this poem the benefit of what all witnesses agree was her amazingly eloquent voice:

THE PILOT OF THE PLAINS

'False,' they said, 'thy Pale-face lover, from the land of waking morn;
Rise and wed thy Redskin wooer, nobler warrior ne'er was born;
Cease thy watching, cease thy dreaming,
    Show the white thine Indian scorn.'

Thus they taunted her, declaring, 'He remembers naught of thee:
Likely some white maid he wooeth, far beyond the inland sea.'
But she answered ever kindly,
    'He will come again to me,'

[7] See 'A Strong Race Opinion on the Indian Girl in Modern Fiction', quoted in Betty Keller, *Pauline: A Biography of Pauline Johnson* (Halifax, 1987), 116–21.

Till the dusk of Indian summer crept athwart the western skies;
But a deeper dusk was burning in her dark and dreaming eyes,
As she scanned the rolling prairie,
Where the foothills fall, and rise.

Till the autumn came and vanished, till the season of the rains,
Till the western world lay fettered in midwinter's crystal chains,
Still she listened for his coming,
Still she watched the distant plains.

Then a night with nor'land tempest, nor'land snows a-swirling fast,
Out upon the pathless prairie came the Pale-face through the blast,
Calling, calling, 'Yakonwita,
I am coming, love, at last.'

Hovered night above, about him, dark its wings and cold and dread;
Never unto trail or tepee were his straying footsteps led;
Till benumbed, he sank, and pillowed
On the drifting snows his head,

Saying, 'O! my Yakonwita call me, call me, be my guide
To the lodge beyond the prairie—for I vowed ere winter died
I would come again, beloved;
I would claim my Indian bride.'

'Yakonwita, Yakonwita!' Oh, the dreariness that strains
Through the voice that calling, quivers, till a whisper but remains,
'Yakonwita, Yakonwita,
I am lost upon the plains.'

But the Silent Spirit hushed him, lulled him as he cried anew,
'Save me, save me! O! beloved, I am Pale but I am true.
Yakonwita, Yakonwita,
I am dying, love, for you.'

Leagues afar, across the prairie, she had risen from her bed,
Roused her kinsmen from their slumber: 'He has come tonight,' she said.
'I can hear him calling, calling;
But his voice is as the dead.'

'Listen!' and they sate all silent, while the tempest louder grew,
And a spirit-voice called faintly, 'I am dying, love, for you.'
Then they wailed, 'O! Yakonwita.
He was Pale but he was true.'

Wrapped she then her ermine round her, stepped without the tepee door,
Saying, 'I must follow, follow, though he call for evermore,
Yakonwita, Yakonwita;'
And they never saw her more.

Late at night, say Indian hunters, when the starlight clouds or wanes,
Far away they see a maiden, misty as the autumn rains,
Guiding with her lamp of moonlight
Hunters lost upon the plains.[8]

Pauline Johnson was a tough and unusual character but, despite her passionate defence of the Indians and especially of Indian women, she had no intention of going off into the forest and living like an Indian of what she called the 'wild'—as opposed to the 'cultivated' variety. Although she was an expert canoeist, she herself was 'cultivated', and preferred Vancouver to the woods.

So what sort of women—if any—*did* end up in the woods, and live to tell the tale—or to have the tale told about them?

It might help to make a division between what we could call the 'first wave'—women of the nineteenth century who were either early settlers themselves or contemporaneous with them—and the 'second wave', women of the twentieth century who followed these first women and either built upon, wrote about, or contrasted their own lives with those of their predecessors.

There is one huge difference between first and second waves: the women of the first wave were not in the North woods of their own volition. They were there because cir-

[8] See 'A Strong Race Opinion on the Indian Girl in Modern Fiction', quoted in Betty Keller, *Pauline: A Biography of Pauline Johnson* (Halifax, 1987), 9.

cumstances and fate—namely their husbands—had dragged them there. None of these women marched off into the woods alone, whereas—as we will see—those of the second wave did.

Three early women writers in Canada offer three different ways in which women of the first wave viewed themselves in relation to the wilderness. (I say 'women writers in Canada' instead of 'Canadian women writers', because none of these women grew up in Canada; but then, neither did a lot of people who have written about the country.) These three are Anna Jameson, Susanna Moodie, and Catherine Parr Traill. None of them wrote fiction or poetry, primarily. Jameson came to Canada because her husband had been appointed Attorney-General in Canada in 1833. She was already a well-known writer, and her Canadian book—*Winter Studies and Summer Rambles in Canada*—was essentially a travel book, the observations of a well-heeled tourist. She did make a two-month excursion into the wilderness, but it was a fully catered excursion, complete with voyageur canoe crew, parasol, and smelling-salts. With others looking after the paddling, cooking, and duck-shooting, she was free to admire the scenery. She was also free to go back to Europe when she got tired of it, and of her husband, which she did.

Susanna Moodie and Catherine Parr Traill were not so free, because they were not so rich. In fact, although they were from the educated class, they were not rich at all. They were sisters, but their versions of a settler's life in a log cabin in the wilderness are quite different. Traill's book, *The Canadian Settler's Guide*, is a practical how-to book for prospective immigrants. She concentrates on coping—recipes, furnishings, and making the best of it. Moodie's book[9] was written as a *warning* to prospective immigrants, especially those of her own class. She emphasizes hardship and catastrophe—people were always stealing things from her or stuffing dead skunks up her

[9] i.e. *Roughing It In The Bush* (1852; repr. London, 1986).

chimney, or the house was catching fire in the middle of the winter. People in her books go mad, commit murder, get lynched; she leans more towards drama and Gothic effects than towards food preparation.

These then are the three patterns: the tourist, the coper, and something we might call 'dismayed'. Although all three women admire the scenery, from a distance—up close it was too full of mosquitoes—and although they comment on the untouched freshness of nature, none of them was that keen on staying in the wilderness. The fact is that unless you had money and could pay for help, being a settler's wife was extremely hard work. And it was hard work of a specific kind. Male protagonists, of the Franklin–Service variety—that is, the romantic variety—are most frequently seen *outdoors*. They are out encountering the land—they penetrate it, they open it up, they stake it out, they grapple with it, they fight with it, they wrest its secrets and its treasures from it, they win or lose. If they stay in their cabins too long they get cabin fever, because they are usually in those cabins by themselves. But what happens in a cabin depends partly on who else, and how many, are in there with you. Women of the first wave were not out on the land, but inside houses with their families, nor do they utilize verbs of the staking and penetrating variety.

A second-wave writer, Margaret Laurence, plays an arpeggio on this theme in her 1974 novel *The Diviners*. Her protagonist, Morag Gunn, lives in a log cabin, but it's a log cabin that's been updated, and has running water and all mod cons. She admires its hand-hewn timbers, but she speculates on the lives of those who hewed them, and especially on the lives of the wives of the hewers:

> You wondered about people like the Cooper family, all those years ago. Trekking in here to take up their homestead. No roads. Bush. Hacking their way. . . . The sheer unthinkable back-and-heartbreaking slog. Women working like horses. Also, probably pregnant most of the time. . . . How many women went mad?

> Loneliness, isolation, strain, despair, overwork, fear. Out there, the bush. In here, a silent worried work-sodden man, squalling brats, an open fireplace, and would the shack catch fire this week or next? In winter, snow up to your thighs. Outdoor privy. People flopping through drifts to the barn to milk the cow. What fun. Healthy life indeed. A wonder they weren't all raving lunatics. Probably many were. It's the full moon, George—Mrs. Cooper always howls like this at such a time—she'll be right as rain come the morning—c'mon there, Sarah, quit crouching in the corner and stop baring your fangs like that—George and me's hungry and would appreciate a spot of grub.[10]

But despite this sort of mordant fun, Morag admires the Coopers, and bestows upon them what is, for her, some very high praise: 'The fact remained that they *had* hacked out a living here. They had *survived*.' And she admires Catherine Parr Traill so much that she uses the woman she refers to as 'Catherine P.T.' as a stick to beat herself with:

> Catherine Parr Traill, one could be quite certain, would not have been found of an early morning sitting over a fourth cup of coffee, mulling, approaching the day in gingerly fashion, trying to size it up. No. No such sloth for Catherine P.T.
>
> *Scene at the Traill Homestead, circa 1840*
>
> CPT out of bed, fully awake, bare feet on the sliver-hazardous floor-boards—no, take that one again. Feet on the homemade hooked rug. Breakfast cooked for the multitude. Out to feed the chickens, stopping briefly on the way back to pull fourteen armloads of weeds out of the vegetable garden and perhaps prune the odd apple tree in passing. The children's education hour, the umpteen little mites lisping enthusiastically over this enlightenment. Cleaning the house, baking two hundred loaves of delicious bread, preserving half a ton of plums, pears, cherries, etcetera. All before lunch. *Catherine Parr Traill, where are you now that we need you? Speak, O lady of blessed memory.*[11]

[10] Margaret Laurence, *The Diviners* (1974; repr. Toronto, 1975), 94, 95. © 1974 by Margaret Laurence.

[11] Ibid. 96.

But Morag doesn't need Traill's canning recipes and hooked-rug patterns. Her own obstacles have to do with her past life and her present relationships. She does not connect with the rich tourism of Anna Jameson, and she has enough Gothicism in her own life so that it's not Susanna Moodie's catastrophes that engage her imagination. What she wants from her pioneer predecessor is something she feels as a lack in herself: she wants to know how to cope.

What had Catherine said, somewhere, about emergencies?
Morag loped over to the bookshelves. . . . Found the pertinent text.

> In cases of emergency, it is folly to fold one's hands and sit down to bewail in abject terror. It is better to be up and doing.[12]

It's the Catherine Parr Traill model—the practical coper—that Joyce Marshall uses for the female protagonist of her story 'The Old Woman'. It dates from the 1950s and features the twentieth century's closest approximation of a settler's wife dragged off into the bush by her husband—a British war-bride. Molly has been brought to northern Québec by her husband, Toddy, the overseer of an electrical power plant connected with a big paper-mill. In transit, Molly thinks of Canada as 'this strange romantic north'—even the name of the town, Missawani, has an exotic ring for her—and the journey to the house by dog-sled echoes Service, and recalls an entire body of northern-landscape imagery:

> It was a long strange journey over the snow, first through pink-streaked grey, then into a sun that first dazzled and then inflamed the eyes. . . . Snow that was flung up coarse and stinging from the feet of the dogs, black brittle fir-trees, birches gleaming like white silk. No sound but the panting breath of the dogs, the dry leather-like squeak of the snow under the sleigh's runners . . .[13]

The house to which Toddy takes Molly is right beside the

[12] Ibid. 97.

[13] Joyce Marshall, from Margaret Atwood and Robert Weaver (eds.), *The Oxford Book of Canadian Short Stories in English* (Toronto, 1986), 93.

waterfall that runs the power generator, and Molly doesn't like the noise of it. Nor does she like the snow, 'blue and treacherous as steel'. Toddy himself loves the sound of the waterfall, but, even more, he loves the power generator, which he refers to as 'my old woman'. He doesn't love much of anything else, including his own sled-dogs—'wolfish brutes'—and his francophone workers—'a lazy bunch of bums'. He doesn't even like spending much time with Molly; although he insists that she stay in the house and not go roaming around, *he's* always down at the generating plant.

Molly rolls up her sleeves, gets busy, and makes a life for herself. She builds up a friendship with the neighbouring folks, through her knack as a midwife. But Toddy becomes more withdrawn, and when Molly suggests that he should take some time off from the plant because it must be difficult being alone with the water always roaring, he explodes in anger.

> 'I have never been bushed . . . How dare you suggest that such a thing could ever happen to me?' . . . Bushed. . . . She was familiar with the term. Toddy used it constantly about others who had come up north to live. He knew the country, but he had been away. And then he had returned alone to this place, where for so long every year the winter buried you, snow-blinded you, the wind screamed up the hill at night, and the water thundered . . .[14]

Things go on, with Molly spending more and more time delivering babies and Toddy spending more and more time with his machine, until finally one evening Molly goes to the power-house to tell Toddy that she's off to another delivery, and finds that he's lost his mind completely.

> . . . his face was quite empty except for a strange glitter that spread from his eyes over his face. He did not answer her. For a moment she forgot that she was not alone with him, until a sound reminded her of Louis-Paul, awake now and standing by the door. And from

[14] Joyce Marshall, from Margaret Atwood and Robert Weaver (eds.), *The Oxford Book of Canadian Short Stories in English* (Toronto, 1986), 96.

the expression of sick shaking terror on his face she knew what the fear had been that she had never allowed herself to name.

'Oh Louis,' she said.

'Come madame,' he said. 'We can do nothing here. In the morning I will take you to Missawani. I will bring the doctor back.'

'But is he safe?' she asked. 'Will he—damage the machines perhaps?'

'Oh no. He would never hurt these machines. For years I watch him fall in love with her. Now she has him for herself.'[15]

The competition for Toddy's soul has been between his wife and 'the old woman'. But what is 'the old woman'? The term itself is ambiguous—it can mean mother, or wife. Or is this a very twisted mother-in-law story in disguise? The tug-of-war between Molly and 'the old woman' is in fact pretty straight Service, as in 'The Lure of the Little Voices',[16] and the last line could be a quotation: 'Now she has him for herself.' It's true that the old woman is a machine and not a landscape, and that it would be possible to read this story as an evils-of-technology allegory. But the old woman is the generator *plus* the power—provided by the waterfall; it's an incarnation of that cold, savage, alluring, female power of the North.

Something interesting starts happening to Canadian female protagonists around the middle of the twentieth century. Instead of going off into the woods to be with a man, they start going off into the woods to be by themselves. And sometimes they're even doing it to get *away* from a man. A good many women's novels have been written in response to the theoretical question *Where did Nora go when she walked out of the doll's house?*, but the solutions have varied according to locale. Sometimes the answer was 'Paris', sometimes—in the 1970s, in the States—it was 'out to get a job'. But one of the answers that has frequently presented itself to Canadian women

[15] Ibid. 102. [16] Service, *Complete Poems*, 23.

writers has been 'off to the woods', or, at the very least, 'off to the summer cottage'.

One of the first to make the trip is Maggie Vardoe, in Ethel Wilson's 1954 novel *Swamp Angel*—not to be confused with Margaret Laurence's 1964 novel *The Stone Angel*, in which the heroine, Hagar Shipley, also takes to the woods, though it's not for long. Hagar is old, and is escaping from her relatives, the nursing-home, and cancer; but Maggie, who is younger, is escaping from her disgusting creep of a husband, Edward Vardoe.

Her main reason for leaving him—apart from his unsuitability in every way—is sexual: she can no longer tolerate her 'outraged endurance of the nights' hateful assaults'. But although she runs out on an individual man, she does not run *towards* a world of women. Instead, she gets a job at a quintessentially male institution, a British Columbia fishing-camp, at a place called Three Loon Lake. She's well suited for this, as she's grown up at a similar fishing-lodge in New Brunswick, and is—like Ethel Wilson—an accomplished fly-fisher herself. But apart from familiarity, what does Three Loon Lake have to offer her?

We get a clue from the description of her first night away from home, in a roadside motel cabin at the edge of the forest.

> As she lay in the dark in the hard double bed and smelled the sweet rough-dried sheets, she saw through the cabin windows the tops of tall firs moving slowly in a small arc, and back, against the starred sky. . . . The place was very still. The only sound was the soft yet potential roar of wind in the trees. The cabin was a safe small world enclosing her.[17]

What Maggie is seeking is refuge, asylum, a spiritual refreshment. The forest, in relation to her, is not female, but neither is it male; it's neuter and also neutral, although alive,

[17] Ethel Wilson, *Swamp Angel* (1954; repr. Toronto, 1990), 34, 35.

and it reflects what the human mind brings to it. For Maggie it's a source of strength: 'Her tormented nights of humiliations between four small walls and in the compass of a double bed were gone, washed away by this air, this freedom, this joy, this singleness and forgetfulness.'[18]

But Maggie is not the only woman at Three Loon Lake. The other one is Vera, wife of the lodge's owner, which makes her Maggie's employer. Vera hates Three Loon Lake. Her associations with the forest, and with isolation, are unhappy, because she grew up on a poor settler-style 'stump farm' with neglectful parents: *her* escape was to the city, but then she made the mistake of marrying a man who, unbeknownst to her, was obsessed with the idea of running a fishing-lodge. So she's been dragged back into the bush against her will, and resents it deeply. She resents Maggie too: although she depends on her, she dislikes Maggie's Catherine Parr Traill-like elbow-grease competence. Finally, in a fit of evil jealousy, she tries to drown herself in the lake. For her the forest is a 'black abyss', the branches of the trees are 'hostile', and the 'singing stinging clarifying wet cold' of the lake—the lake Maggie swims in so blissfully—gives her pneumonia. For Ethel Wilson, as for Gwen MacEwen, the mind of the observing individual helps to determine what is observed. Maggie's wilderness-life is chosen; she is a free agent, and Vera is not; this is what makes the difference.

But Maggie Vardoe doesn't go quite all the way. Although she's alone inside her cabin at Three Loon Lake, she's still connected with other people, and sees her role in relation to them: she stays on because she's become attached to Vera's young son Alan, and because she feels a moral obligation to nurse Vera back to emotional and physical health (although this may not work, because unhappy superbitch Vera is close to being a lost cause). The wilderness, then, gives Maggie strength, but at a mediated distance.

[18] Ibid. 122.

Marian Engel, on the other hand, takes her heroine beyond the bounds of the human, close to the heart of the natural Other, in her 1976 novel *Bear.* Robert Kroetsch's protagonist Jeremy Sadness falls in love with a woman dressed up in a bear-suit, but Engel's character, Lou, falls in love with the real thing: an actual bear. (Incidentally, you don't call anyone 'Lou' in Canada without a sly sideways glance at Robert Service's 'lady that's known as Lou'. Like Service's Lou, Engel's is no lady.)

Here is how it comes about:

Lou is a bibliographer who works at a place in Toronto called the Historical Institute. All we're told about it is that it's full of old stuff, such as 'a signed photograph of the founder of a seed company absorbed by a rival competitor', and that no one ever leaves it anything of real importance. But finally someone does, maybe: the Institute is left a whole estate, situated on an island near what I can only deduce must be the northern shore of Lake Huron, and Lou is sent up there by herself to catalogue the contents of the house and advise about whether or not the thing can be turned into a Summer Institute.

On her way to the island, Lou discovers that she's been here before. 'She remembered a beach, a lake the colour of silver, something sad happening. Something, yes, that happened when she was very young, some loss. It struck her as strange that she had never come back to this part of the world.'[19] We come to suspect that the 'something sad' was in fact her departure, and that the loss was the loss of the natural world itself. Almost immediately she has the sense of being 'reborn'.

When she reaches the house itself, which isn't a rationalist rectangle but a classic Fowler's octagon, she finds that it contains not only an intact nineteenth-century library—with most of the books imported from England—but also, living in the original settler's log cabin out behind the house, a bear.

[19] Marion Engel, *Bear* (Toronto, 1976), 19.

The bear is there because there has always been a bear there. The builder of the house, an eccentric, romantically minded Englishman named Colonel Cary, knew and admired Byron, the bear-keeping English poet; or this is the explanation we are given. But the person who knows the most about the bear, and has an affinity with it, is a 100-year-old Indian woman named Lucy. (Shades of Wordsworth's nature girl? Probably, since it's Engel doing the writing. Lucy doesn't look 100, we are told: 'only eternal'.) Lucy tells Lou two things: that the bear is a good bear, and that there's a way of winning it over: 'Shit with the bear,' she said. 'He like you, then. Morning, you shit, he shit. Bear lives by smell. He like you.'[20] Having delivered herself of this Jungian wise-old-woman magic charm, Lucy goes off, leaving Lou to cope with the bear.

Lou's first response to her situation is an enormous sense of freedom and joy. She does not make the mistake of scorning Indian gifts. She follows Lucy's advice, and finds that it works—she can take the bear tamely out for walks, she can take it swimming in the lake with her. The bear—which has looked mangy, old, and dirty when she's first seen him—begins to return to life. One night, while she's researching Colonel Cary and his contemporary Beau Brummel, she forgets to chain up the bear and it comes upstairs and lies down in front of the fireplace, like a dog, and she realizes it knows its way around the house.

'She . . . kicked off her shoes, and found herself running her bare foot over his thick, soft coat, exploring it with her toes, finding it had depths and depths, layers and layers. . . . She looked up at Cary and down at the bear and was suddenly exquisitely happy. Worlds changed. Two men in scarlet uniforms. . . . She felt victorious over them; she felt she was their inheritor: a woman rubbing her foot in

[20] Ibid. 49.

the thick black pelt of a bear was more than they could have imagined. More, too, than a military victory: splendour.[21]

She does not conquer the natural world, or penetrate it—she befriends it.

One thing leads to another, and Lou discovers by accident that the bear has an amazing and unself-conscious and undemanding talent for oral sex. Not only that, he's better at it than any man she's ever known. Not only that, he's better than any man she's ever known in almost every way. In particular, he doesn't make her feel inadequate or dirty. We get the details of some of her affairs with men, and they are not happy: the Director of the Institute, for instance, makes love to her on top of the antique maps and the antique desk at the office, but she knows 'in her heart that what he wanted was not her waning flesh but elegant eighteenth-century keyholes, of which there is a shortage in Ontario'.

She tries to understand the bear, through old bits of mythology, scraps of which have been left helpfully tucked into the books by the first Colonel Cary. She tries to figure out what the bear thinks. She becomes more and more obsessed by the bear, and begins projecting all kinds of identities on to him. Finally, she wants his identity from him: 'Bear, make me comfortable in the world at last. Give me your skin.' She tries dancing with him to Greek music on the radio. Not a great success. She eats with him, down on the floor on all fours. She realizes she is going definitely bush—'Her hair and her eyes were wild. Her skin was brown and her body was different and her face was not the same face she had seen before'—and she cleans herself up and goes to the mainland and makes love with the only person she knows there, a man called Homer who's in charge of the house in the winters: but Homer does absolutely nothing for her. It's also Homer's opinion that the bear is just a bear, nothing more, and that it's also a wild

[21] Marion Engel, *Bear* (Toronto, 1976), 57.

animal really, and should be treated with caution; but Lou doesn't want to hear this. She wants to dress up in strange fur garments and stay with the bear for ever and ever, and keep him safe from harm. She loves him, and she believes he loves her.

But, alas, Homer is right, at least on one plane of existence. The bear is a bear, and when Lou tries to be a bear too—when she tries to get the bear to mate with her as if she were a female bear—the bear gives her a rebuking swipe with his paw and tears half the skin off her back. But this does not turn into a story of Nature Red in Tooth and Claw. The bear does not become savage, he does not destroy Lou. In fact, Lou is rather cheered up by the big bear-mark she will now always have on her skin. It's a sort of initiatory tattoo.

Women-in-the-wilderness books frequently contain a mirror scene—a scene in which the woman looks in the mirror and sees that she has been altered. Susanna Moodie, for instance, after seven years in the bush, remarks that she has been given the face of an old woman. But Lou, in *her* mirror scene, notes on herself, 'She was different. She seemed to have the body of a much younger woman.' She gets another gift from the bear, too: 'What had passed to her from him she did not know. Certainly it was not the seed of heroes, or magic, or any astounding virtue, for she continued to be herself. But for one strange, sharp moment she could feel in her pores and the taste of her own mouth that she knew what the world was for. She felt not that she was at last human, but that she was at last clean.'[22]

Things return to themselves, the bear becomes the enigma it has always been, and goes off with Lucy in a motorboat, metamorphosed from ultra-male to 'a fat dignified old woman with his nose to the wind', and Lou returns to the city, taking with her the one thing she's wanted to rescue from the house full of English books—a first edition of our old friend

[22] Ibid. 136, 137.

Richardson's *Wacousta.* What her wilderness sojourn has given Lou is an authentic, strong self; she's recovered the thing lost, she's got rid of her alienated sadness. And the bear, in its transcendent aspect, is still with her as a guardian spirit: As she drives south, 'It was a brilliant night, all star-shine, and overhead the Great Bear and his thirty-seven thousand virgins kept her company.'

In the later 1970s and 1980s, the patter of female feet heading for the hills, not to mention the valleys, islands, forests, and plains, becomes a small stampede. One could mention Aretha Van Herk's *Tent Peg,* in which a woman disguises herself as a man and goes off on that modern version of prospecting, a geological expedition; or Joan Barfoot's *Abra,* in which another awful man is discarded; or Susan Swan's *Last of the Golden Girls,* which takes place in summer-cottage country. And for a very recent and startling view of a woman in nineteenth-century frontier society, see Alice Munro's complex story 'Meneseteung', in the collection *Friend of My Youth.*

It's difficult to generalize, and there's no such thing as an always consistent, always the same Woman's Voice, with capital letters, any more than there's a single Man's Voice; but I'll generalize a little bit anyway. Frequently, then, when second-wave women write about the wilderness, they render it female in relation to male characters, as MacEwen does in both her Franklin and her Grey Owl poems, and as Joyce Marshall does in her Old Woman story. But when the protagonist is a woman, the wilderness is apt to be sexually neuter. It's also apt to be refreshing or renewing in some way, and in the 1970s at any rate this renewal has something to do with the absence of men from the scene. A more or less feminist enterprise gets entwined with the older Seton-and-Grey-Owl project of turning the wilderness into a spiritual health spa.

In my last lecture, I promised you a female Wendigo, and here it is. As you'll recall, the Wendigo is a terrifying, cannibalistic monster of the Algonquian forests. But, although oral legends tell us that a Wendigo may be either male or female, all the written ones we talked about were male.

The Wendigo appears in its—to my knowledge—first literary female incarnation in Ann Tracy's 1990 novel, called, with stunning aptness, *Winter Hunger*. It may or may not be true that we live each event twice, once as tragedy and once again as farce, but its' certainly true in literature, and in *Winter Hunger* the Wendigo motif, such a focus of horror and/or spiritual dismay in its previously studied versions, is used by the author for subversive and often hilarious purposes.

The Wendigo belief, in this book, is front and centre; in fact, it *is* the centre. The story is told from the point of view of a white, originally American student of anthropology named Alan Hooper, yet another avatar of the silly-fool graduate student we last saw done up in the braids and fringes of Robert Kroetsch's Jeremy Sadness. Alan has been studying at the University of Toronto, and is now living in a claustrophobia-inducing trailer beside a remote northern lake, writing his thesis on food-gathering patterns in three Native communities. With him live his aloof, beautiful wife, Diana, whom Alan adores with an obsessive passion, and their infant son Cam, whom Alan does not very much like, because he takes up too much of Diana's time and also too much of her chest area. Names are significant in this book, and I would remind you—as the author does, fairly early on—of 'Diana's' association with hunting. As for 'Cam', it's my own theory, unsubstantiated by any documentary evidence, that it was chosen to rhyme with 'Spam', which is frequently eaten in the North.

The family's trailer is located beside Wino Day Lake, which, the Indians insist, is wrong on the white-man's maps. Alan doesn't get it at first, but a better-educated person would have known what to make of a word beginning with W and containing the consonants N and D and the vowels I and O. In fact, a better-educated person would have high-tailed it out of Wino Day Lake as fast as possible, and *certainly* after the Native men have announced that they feel 'something hungry' watching them out of the forest, and have remarked that this only happens sometimes in the winter.

The presiding symbol-maker at Wino Day is Proxene Ratfat, the local bag-lady, who specializes in abstract drawings of elusive shapes with 'little fanged edgings like rickrack gone bad'. Alan's wife, Diana, is collecting these with an eye to writing something about them. Diana has also made loving friends with an old woman called Naomi, of biblical resonance. Naomi starts telling Diana some Wendigo stories, in particular one about her own great-grandmother, who was stranged and burned up by her relatives when it was thought she was turning into a Wendigo. Alan is sceptical about these stories—scepticism in such matters is a bad mistake, as we've noted—whereas Diana is enthusiastic about them. For Diana, the world is 'dense with possibility, both magical and horrific', because Diana, although aloof, is 'receptive'; she's 'made of steel and crystal inside—a good receiver, a good transmitter, but not what you might call pliable'.

As the winter wears on, Naomi becomes apathetic and won't eat. Finally she announces that she would like to eat Diana. In order to melt the Wendigo heart of ice growing within her, her relatives fill her full of boiling-hot tea, using a funnel; the operation is successful, and she dies. Alan is horrified by what he sees as cruelty due to superstition; Diana, on the other hand, is heart-broken, because she's lost her friend and mentor.

Shortly after this—shortly, in fact, after he trips in the winter darkness and falls face-down into the intestines of a man who's been cut open in a knife-fight—Alan starts to fantasize about eating Diana himself. His day-dreaming begins with a longing for a closer union with her—there has always been an inner, a 'quintessential' Diana he's been unable to possess—and he's preoccupied by the 'greed to have all of her'. He tries sex, longing for 'a magical interpenetration', but it's just not enough:

He felt for the first time that his penis was constructed on the wrong principle. . . . Now this heretofore inspired piece of anatomical engineering was useless to his purposes, for it only put out, and he needed something that would draw in; he needed it to work like the hose of an enormously powerful industrial vacuum cleaner, pulling Diana into his body. If he put it in her and turned it on, could she not be sucked inside out like a pink silk stocking and rearranged under his own skin, stretched so that her fingers fit inside his fingers, her knees cupped into his own, and her nipples charged his from behind like the positive ends of two dry cell batteries?[23]

His interest in eating Diana expands to become an interest in eating almost everyone else. Along with these werewolfish fantasies, he has to contend with the repugnance he feels for any kind of real food, and he begins to starve. Finally, he has a nightmare in which he turns into one of Proxene Ratfat's toothy drawings and bites Diana in his sleep, after which he has a hallucinatory vision of Cam as a roast suckling pig. Horrified, he purchases a plane ticket to Toronto, and escapes from Wino Day Lake in order to cure himself of what he still regards a bad case of bushy cabin fever.

In the process of weaning himself back on to the real food he's been finding so repulsive, he spends some time in the University of Toronto library, where he comes across—guess what!—Morton W. Teicher's article on windigo psychosis. (My own theory—based on internal evidence—is that he comes across John Robert Colombo's *Windigo* book.) At first this is reassuring: if belief determines behaviour and he isn't a believer, then he can't be turning into a Wendigo. But then he realizes *why* the Indians say that the name of Wino Day Lake is wrong on the map: it should be Wendigo Lake. He's left his wife and child surrounded by a whole village-full of true

[23] Ann Tracy, *Winter Hunger* (Fredericton, New Brunswick, 1990), 101.

believers. Given the chance, Naomi might very well have eaten Diana. Alan is alarmed: 'Diana was *his* wife. And while he was very glad not to have eaten her himself in a rash moment, he was damned if anybody else was going to commit that kind of gastronomical rape in his stead.'

In a panic, he rushes back to Wino Day Lake, only to find that Diana is not there to meet him at the plane. He hurries off to his trailer, vowing to 'never let Diana out of his sight again. . . . He would go back and feed off [her] sweetness for the rest of his life. . . . His craziness had been after all merely a nervous and misguided manifestation of his great need of her nourishing presence.'

He finds Diana sitting in the kitchen, stark naked. 'How very odd,' he thinks, 'and it didn't seem like a welcoming gesture designed for him, for she hadn't combed her hair, perhaps not for a long time, and her expression—well, "self-contained" hardly touched it, and "haughty" was too out-going.' He realizes that Diana hasn't even washed for some time. And something, or someone, is missing.

> 'Where is Cam?' he said distinctly.
>
> The change from gabble seemed to catch Diana's attention. 'Gone,' she said indifferently. Then a jaunty and feral grin animated her face and she added, '—almost.'[24]

Cam, or what is left of Cam—is on a platter on the kitchen counter. Alan feels 'a wave of real grief . . . fellow-feeling for a lover whose source of nourishment had turned against him'. But he chooses to believe that Diana is merely having a nervous breakdown, that they can go away together and she will be not only cured, but bound even more closely to him by the awful secret between them.

As he reaches out his arms to embrace Diana, she embraces him, too, but with a hunting-knife concealed behind her back.

[24] Ann Tracy, *Winter Hunger* (Fredericton, New Brunswick, 1990), 163.

> Time stopped as his mind protested at the wrongness of it—no, no, this was not fair. He did not want them to be one if it meant her absorption of him. . . . This was not love, not romance, it was pain and gastric juices and an ignominious end in the intestines—too intimate by half.[25]

His last fantasy is 'a vision of his death à la Ratfat: two small human figures (one very small indeed) sucked into a monstrous, fang-fringed orifice'.

Has the unfortunate Alan's dream of closer-than-close intimacy been picked up by Diana, the crystalline good receiver, the good transmitter, and been played back to him? Or has the too-impressionable Diana, during a bad case of cabin fever, been taken over by the Wendigo mythology? Or are Proxene Ratfat and the other members of the Native community right, and can people really turn into Wendigoes? Whatever you choose to believe, the result—for Alan—is the same.

But what *Winter Hunger* is really out to puncture is the concept of romantic love. The greed behind Alan's hunger is his stifling wish to possess Diana, to possess *all* of her, especially her soul. He is devoured, in the end, by his own too-intense, too-selfish desires; and once again the Wendigo becomes the incarnation of spiritual selfishness, though this time with a sexual twist.

For the student of Wendigoes, the interesting factor is that all the Wendigoes described in the story are not male—as they usually are—but female: Naomi's grandmother, Naomi herself, Diana. Even the stylized Wendigoes drawn by Proxene Ratfat resemble those schematic *vagina dentata* drawings in anthropology textbooks; and it's only the *women* of the community who collect them, presumably because it's only the women who understand them. The Wendigo, terrifying male monster and ravenous personification of winter and of

[25] Ibid. 165.

scarcity, has been turned inside out and has become an enormous and ravenous female body part.

In my end is my beginning; or at least, in the end of this lecture is the beginning of the first one. Is Diana the victim of the North—or is she, as seems more likely, its incarnation? As I've said, there is a tendency for women to write the wilderness as female when it's a question of a male protagonist, and this is proof with a vengeance. 'Long have I waited lonely, shunned as a thing accurst,' says Robert Service's female Yukon; just as Diana says, in her cool, icy, crystalline voice, 'I've been waiting for you.' Here is Gwendolyn MacEwen's 'giant virginal strait of Victoria' crushing men in her 'stubborn loins, her horrible house, her white asylum in an ugly marriage'. Only this time it's, well, funny. Though like Proxene Ratfat's drawings, which are understood only by women, perhaps it's funny only to females.

This has been only a small sampler of Canadian Northern imagery, but I hope it has served to give you an indication, not only of the nature of some of these image-clusters, but of their longevity, of the varieties of ways in which they may be used by writers, and of their vitality. Writers, evidently, continue to be drawn to them. If you would like another perspective on Northernness in Canadian art, check the catalogue for the exhibition which was at the Barbican in London, entitled 'The True North'.[26] It's worth noting that many of the painters represented were inspired by earlier poets and writers, and that they themselves have inspired others in their turn.

Now, here's a bit the Canadianists and thesis-writers in the audience have been waiting for. A number of you have asked me, since I've been here, whether any of this stuff is connected at all with my own work. The answer, in a word, is Yes.

[26] Michael Tooby (ed.), *The True North: Canadian Landscape Painting 1896–1939* (London, 1991).

Writers, being highly self-involved people, would be unlikely to give lectures on anything that wasn't connected in some way with their own work, and I am no exception. My own woman-in-the-woods novel was published in 1972, and is called *Surfacing*. My rewriting of Susanna Moodie is in a poem-sequence entitled 'The Journals of Susanna Moodie', now available here in *Poems 1965–1975*, in handy Virago paperback. Franklin and, more especially, the three permafrozen sailors from his expedition appear in a story called 'The Age of Lead'—as you may guess, I was interested in the lead-poisoning aspect—and Wacousta turns up as the name of a summer house in a story called 'Wilderness Tips'. A man who longs to be an Indian is to be found in the same story. These stories themselves are in a collection also called *Wilderness Tips*, in which—Yes!—the *Titanic* metaphorically sinks, taking all of us down with it. (In fact, this important vessel gets into not one but *two* stories.)

Lecture series are apt to end with a conclusion which tends to be admonitory or moralistic, and mine will follow convention and leave you with a disturbing thought. Next time you're in Pearson International airport in Toronto, wander through the souvenir shops and look at what is to be found there in the way of instant national identity. Chances are you will find a light-up Mountie or two, some maple leaves made out of maple sugar, many notecards with birds, animals, and landscapes on them, some Eskimo carvings, some smoked salmon in wooden boxes with Haida motifs, and a lot of mittens. You will also find many gift books with names like *Beautiful Canada*, and these will have a large supply of Northern scenes—scenes shown as vast, empty, untouched, luminous, numinous, pristine, and endless. Canadians have long taken the North for granted, and we've invested a large percentage of our feelings about identity and belonging in it.

But the bad news is coming in: the North is not endless. It is not vast and strong, and capable of devouring and digesting all the human dirt thrown its way. The holes in the ozone layer are getting bigger every year; the forest, when you fly over it in a plane, shows enormous wastelands of stumps; erosion, pollution, and ruthless exploitation are taking their toll.

The edifice of Northern imagery we've been discussing in these lectures was erected on a reality; if that reality ceases to exist, the imagery, too, will cease to have any resonance or meaning, except as a sort of indecipherable hieroglyphic. The North will be neither female nor male, neither fearful nor health-giving, because it will be dead. The earth, like trees, dies from the top down. The things that are killing the North will kill, if left unchecked, everything else.

BIBLIOGRAPHY

ATWOOD, MARGARET, *Poems 1965–1975* (London: Virago, 1991).

—— *Surfacing*, Virago Modern Classic no. 8 (London: Virago, 1979).

—— *Wilderness Tips* (London: Virago, 1992).

—— (ed.), *The New Oxford Book of Canadian Verse in English* (Toronto: Oxford University Press, 1982).

—— and ROBERT WEAVER (eds.), *The Oxford Book of Canadian Short Stories in English* (Toronto: Oxford University Press, 1986).

BARFOOT, JOAN, *Abra* (Toronto: McGraw-Hill Ryerson, 1978).

BEASLEY, DAVID R., *The Canadian Don Quixote: The Life and Works of Major John Richardson Canada's First Novelist* (Erin, Ont.: Porcupine's Quill, 1977).

BEATTIE, OWEN, and JOHN GEIGER, *Frozen in Time* (Saskatoon: Western Producer Prairie Books, 1989).

BIRNEY, EARLE, *The Collected Poems of Earle Birney*, 2 vols. (Toronto: McClelland & Stewart, 1975).

BLAKE, W. H., *Brown Waters and Other Sketches* (Toronto: Macmillan, 1940).

BOWERING, GEORGE, *The Gangs of Kosmos* (Toronto: House of Anansi, 1969).

COLOMBO, JOHN ROBERT (ed.), *Windigo: An Anthology of Fact and Fantastic Fiction* (Saskatoon: Western Producer Prairie Books, 1982).

DAVIES, ROBERTSON, *The Manticore* (Toronto: Macmillan, 1972).

DICKSON, LOVAT, *Wilderness Man: The Strange Story of Grey Owl* (Toronto: Macmillan, 1973).

DREW, WAYLAND, *The Wabeno Feast* (Toronto: House of Anansi, 1973).

DRUMMOND, WILLIAM HENRY, *Dr W. H. Drummond's Complete Poems* (Toronto: McClelland & Stewart, 1926).

ENGEL, MARION, *Bear* (Toronto: McClelland & Stewart, 1976).

FRANKLIN, JOHN, *Narrative of a Journey to the Shores of the Polar Sea in the Years 1819, 1820, 1821, and 1822* (1823; repr. Rutland, Vt.: Tuttle, 1970).

GIBSON, GRAEME, *Perpetual Motion* (Toronto: McClelland & Stewart, 1982).

GREY OWL (WA-SHA-QUON-ASIN), *The Men of the Last Frontier*, Macmillan Paperback no. 42 (Toronto: Macmillan, 1989).

—— *Pilgrims of the Wild* (1935; repr. Toronto: Macmillan, 1970).

GRIFFITHS, LINDA, and MARIA CAMPBELL, *The Book of Jessica: A Theatrical Transformation* (Toronto: Coach House, 1989).

GUDGEON, CHRIS, *An Unfinished Conversation: The Life and Music of Stan Rogers* (Toronto: Viking/Penguin, 1993).

HEMSWORTH, WADE, *The Songs of Wade Hemsworth*, (ed.) Hugh Verrier (Waterloo, Ont.: Penumbra, 1990).

HIGHWAY, TOMSON, *Dry Lips Oughta Move to Kapuskasing* (Saskatoon: Fifth House, 1989).

—— *The Rez Sisters* (Saskatoon: Fifth House, 1988).

HURLEY, MICHAEL, *The Borders of Nightmare: The Fiction of John Richardson* (Toronto: University of Toronto Press, 1992).

JAMESON, ANNA BROWNELL, *Winter Studies and Summer Rambles in Canada*, New Canadian Library no. 46 (1838; repr. Toronto: McClelland & Stewart, 1985).

JILES, PAULETTE, *Celestial Navigation: Poems* (Toronto: McClelland & Stewart, 1984).

JOHNSON, E. PAULINE, *Flint and Feather: The Complete Poems of E. Pauline Johnson (Tekahionwake)* (Toronto: Hodder & Stoughton, 1969).

—— *Legends of Vancouver* (Toronto: McClelland & Stewart, 1961).

KELLER, BETTY, *Black Wolf: The Life of Ernest Thompson Seton* (Vancouver: Douglas & McIntyre, 1984).

—— *Pauline: A Biography of Pauline Johnson* (Halifax: Goodread Biographies, 1987).

KELLY, M. T., *Breath Dances Between Them* (Toronto: Stoddart, 1991).

KING, THOMAS, *All My Relations: An Anthology of Contemporary Canadian Native Fiction* (Toronto: McClelland & Stewart, 1990).

KLEIN, A. M., *The Rocking Chair and Other Poems* (Toronto: Ryerson, 1948).

KROETSCH, ROBERT, *Gone Indian* (Toronto: New Press, 1973).

Bibliography

LAURENCE, MARGARET, *The Diviners* (1974; repr. Toronto: Bantam, 1975).

—— *The Stone Angel* (Toronto: McClelland & Stewart, 1968).

LEACOCK, STEPHEN, *Literary Lapses*, New Canadian Library no. 3 (Toronto: McClelland & Stewart, 1971).

MACEWEN, GWENDOLYN, *Afterworlds* (Toronto: McClelland & Stewart, 1987).

MCGREGOR, GAILE, *The Wacousta Syndrome: Explorations in the Canadian Landscape* (Toronto: University of Toronto Press, 1985).

MAIR, CHARLES, *Tecumseh, a Drama, and Canadian Poems*, Master-Works of Canadian Authors, xiv, (ed.) John W. Garvin (Toronto: Radisson Society of Canada, 1926).

MANGUEL, ALBERTO (ed.), *The Oxford Book of Canadian Ghost Stories* (Toronto: Oxford University Press, 1990).

MITCHAM, ALISON, *The Northern Imagination: A Study of Northern Canadian Literature* (Moonbeam, Ont.: Penumbra, 1983).

MOODIE, SUSANNA, *Roughing It In The Bush* (1852; repr. London: Virago, 1986).

MUNRO, ALICE, *Dance of the Happy Shades* (Toronto: Ryerson, 1968).

—— *Friend of My Youth* (Toronto: McClelland & Stewart, 1990).

—— *Lives of Girls & Women* (Toronto: McGraw-Hill Ryerson, 1971).

NASH, OGDEN, *Bed Riddance: A Posy for the Indisposed* (Boston: Little, Brown, 1969).

NEWLOVE, JOHN, *The Fatman: Selected Poems 1962–1972* (Toronto: McClelland & Stewart, 1977).

PITT, DAVID G., *E. J. Pratt: The Master Years, 1927–1964* (Toronto: University of Toronto Press, 1987).

POLK, JAMES, *Wilderness Writers* (Toronto: Clarke, Irwin, 1972).

PRATT, E. J., *Complete Poems*, parts 1 and 2, (ed.) Sandra Djwa and R. G. Moyles (Toronto: University of Toronto Press, 1989).

PURDY, AL, *The Collected Poems of Al Purdy*, (ed.) Russell Brown (Toronto: McClelland & Stewart, 1986).

REANEY, JAMES, *Wacousta!* (Toronto: Press Porcépic, 1979).

RICHARDSON, JOHN, *The Canadian Brothers: or, The Prophecy Fulfilled: A Tale of the Late American War* (Toronto: University of Toronto Press, 1976).

RICHARDSON, JOHN, *Eight Years in Canada* (New York: S. R., Johnson Reprint Corporation, 1967).

—— *Wacousta; or, The Prophecy*, New Canadian Library no. 58 (abridged) (Toronto: McClelland & Stewart, 1988).

RICHLER, MORDECAI, *Solomon Gursky Was Here* (Markham, Ont.: Viking, 1989).

ROSS, W. GILLIES, *Arctic Whalers, Icy Seas: Narratives of the Davis Strait Whale Fishery* (Toronto: Irwin, 1985).

SERVICE, ROBERT W., *The Complete Poems of Robert Service* (New York: Dodd Mead, 1945).

SETON, ERNEST THOMPSON, *The Book of Woodcraft* (1912; repr. Berkeley: Creative Arts Books, 1988).

—— *Two Little Savages: Being the Adventures of Two Boys who Lived as Indians and what they Learned* (Garden City, NY: Doubleday, 1959).

SMITH, A. J. M. (ed.), *The Book of Canadian Poetry: A Critical and Historical Anthology*, 3rd edn. (Toronto: W. J. Gage, 1957).

STAINES, DAVID (ed.), *The Canadian Imagination: Dimensions of a Literary Culture* (Cambridge, Mass.: Harvard University Press, 1977).

SWAN, SUSAN, *The Last of the Golden Girls* (Toronto: Lester Orpen Dennis, 1989).

TOOBY, MICHAEL (ed.), *The Truth North: Canadian Landscape Painting 1896–1939* (London: Lund Humphries in association with the Barbican Art Gallery, 1991).

TRACY, ANN, *Winter Hunger* (Fredericton, New Brunswick: Goose Lane Editions, 1990).

TRAILL, CATHERINE PARR, *The Canadian Settler's Guide*, New Canadian Library no. 64 (1855; repr. Toronto: McClelland & Stewart, 1969).

VAN HERK, ARITHA, *Tent Peg* (Toronto: McClelland & Stewart, 1981).

WHALLEY, GEORGE, *the Legend of John Hornby* (London: Murray, 1962).

WIEBE, RUDY, *A Discovery of Strangers* (Toronto: Alfred A. Knopf, 1994).

—— *The Mad Trapper* (Toronto: McClelland & Stewart, 1980).

—— *Playing Dead: A Contemplation Concerning the Arctic* (Edmonton: NeWest, 1989).

—— *Where is the Voice Coming From?* (Toronto: McClelland & Stewart, 1974).

Wilson, Ethel, *Swamp Angel* (1954; repr. Toronto: McClelland & Stewart, 1990).

PERMISSIONS

Grateful acknowledgement is made to the following for permission to reprint previously published material:

George Bowering: Excerpts from the poem 'Windigo' reprinted from *The Gangs of Kosmos*, © 1969, George Bowering, reprinted by kind permission of the author.

Chatto & Windus: Excerpts from *Solomon Gursky Was Here* by Mordecai Richler, © 1989, Mordecai Richler, reprinted by permission of Chatto & Windus.

Curtis Brown Ltd: 'The Wendigo' by Ogden Nash, from *Verses from 1929 On*, © 1952, Ogden Nash, renewed, reprinted by permission of Curtis Brown Ltd.

Fogarty's Cove Music and D. Ariel Rogers: 'Northwest Passage' by Stan Rogers, © 1981, Fogarty's Cove Music, reprinted by permission of Fogarty's Cove Music and D. Ariel Rogers.

Goose Lane Editions: Excerpts from *Winter Hunger* by Ann Tracy, © 1990, Ann Tracy, reprinted by permission of Goose Lane Editions.

Graeme Gibson: Excerpts from *Perpetual Motion* by Graeme Gibson, © 1982, Graeme Gibson, reprinted by kind permission of the author.

Paulette Jiles: 'Windigo' from *Celestial Navigation: Poems* by Paulette Jiles, © 1984, Paulette Jiles, reprinted by kind permission of the author.

Little, Brown and Company: 'The Wendigo' from *Verses from 1929 On* by Ogden Nash, © 1953, Ogden Nash, reprinted by permission of Little, Brown and Company.

Permissions

McClelland & Stewart: The poem 'Can. Lit.' from *Collected Poems* by Earle Birney, © 1977, Earle Birney, used by permission of The Canadian Publishers, McClelland & Stewart, Toronto. Excerpts from *Perpetual Motion* by Graeme Gibson, © 1982, Graeme Gibson, used by permission of The Canadian Publishers, McClelland & Stewart, Toronto. Excerpts from *Bear* by Marion Engel, © 1976, Marion Engel, used by permission of The Canadian Publishers, McClelland & Stewart, Toronto. The poems 'The Names', 'Grey Owl's Poem', and excerpts from the poem 'Terror & Erebus' from *Afterworlds* by Gwendolyn MacEwen, © 1987, Gwendolyn MacEwen, used by permission of The Canadian Publishers, McClelland & Stewart, Toronto. The poem 'The Pride' from *The Fatman* by John Newlove, © 1977, John Newlove, used by permission of The Canadian Publishers, McClelland & Stewart, Toronto. Excerpt from the poem 'The Northwest Passage' by Al Purdy from *The Collected Poems of Al Purdy*, © 1986, Al Burdy, used by permission of The Canadian Publishers, McClelland & Stewart, Toronto.

Macmillan Canada: Excerpts from *Swamp Angel* by Ethel Wilson, © 1954, reprinted by permission of Macmillan Canada.

New End Inc.: Excerpts from *The Diviners* by Margaret Laurence, © 1974, Margaret Laurence, reprinted by permission of New End Inc.

Penguin Books Canada: Excerpts from *Solomon Gursky Was Here* by Mordecai Richler, © 1989, Mordecai Richler, reprinted by permission of Penguin Books Canada Limited.

Penumbra Press: Excerpts from *The Songs of Wade Hemsworth* ed. Hugh Verrier, © 1990, Wade Hemsworth/Penumbra Press/Hugh Verrier, published with the permission of Penumbra Press.

Random House Inc.: Excerpts from *Solomon Gursky Was Here* by Mordecai Richler, © 1989, Mordecai Richler, reprinted by permission of Random House Inc.

Permissions

James Reaney: Excerpts from *Wacousta!* by James Reaney, © 1979, James Reaney, reprinted by kind permission of the author.

Russell & Volkening Inc.: Excerpts from *Bear* by Marion Engel, © 1976, Marion Engel, reprinted by permission of Russell & Volkening, Inc.

Stoddart Publishing Co. Limited: Excerpts from 'Case Histories' from *Breath Dances Between Them* by M. T. Kelly, © 1991, M. T. Kelly, reprinted with the permission of Stoddart Publishing Co Limited, Don Mills, Ont. Excerpts from *Gone Indian* by Robert Kroetsch, © 1973, Robert Kroetsch, reprinted with the permission of Stoddart Publishing Co Limited, Don Mills, Ont. Excerpts from The *Wabeno Feast* by Wayland Drew, © 1973, Wayland Drew, reprinted with the permission of Stoddart Publishing Co Limited, Don Mills, Ont.

University of Toronto Press: Excerpts from 'The Titanic' by E. J. Pratt from *E. J. Pratt, Complete Poems*, © 1989, University of Toronto Press, reprinted with the permission of the publisher.

Quotations from *Afterworlds* by Gwendolyn MacEwen, © 1987, Gwendolyn MacEwen, are reprinted by kind permission of the author's family.

INDEX

Index